Cambridge Elements ≡

Elements in the Philosophy of Science
edited by
Jacob Stegenga
University of Cambridge

SCIENTIFIC
REPRESENTATION

James Nguyen
Stockholm University

Roman Frigg
London School of Economics and Political Science

CAMBRIDGE
UNIVERSITY PRESS

CAMBRIDGE
UNIVERSITY PRESS

University Printing House, Cambridge CB2 8BS, United Kingdom

One Liberty Plaza, 20th Floor, New York, NY 10006, USA

477 Williamstown Road, Port Melbourne, VIC 3207, Australia

314–321, 3rd Floor, Plot 3, Splendor Forum, Jasola District Centre,
New Delhi – 110025, India

103 Penang Road, #05–06/07, Visioncrest Commercial, Singapore 238467

Cambridge University Press is part of the University of Cambridge.

It furthers the University's mission by disseminating knowledge in the pursuit of education, learning, and research at the highest international levels of excellence.

www.cambridge.org
Information on this title: www.cambridge.org/9781009009157
DOI: 10.1017/9781009003575

When citing this work, please include a reference to the DOI 10.1017/9781009003575

First published 2022

A catalogue record for this publication is available from the British Library.

ISBN 978-1-009-00915-7 Paperback
ISSN 2517-7273 (online)
ISSN 2517-7265 (print)

Cambridge University Press has no responsibility for the persistence or accuracy of URLs for external or third-party internet websites referred to in this publication and does not guarantee that any content on such websites is, or will remain, accurate or appropriate.

Scientific Representation

Elements in the Philosophy of Science

DOI: 10.1017/9781009003575
First published online: August 2022

James Nguyen
Stockholm University

Roman Frigg
London School of Economics and Political Science

Author for correspondence: James Nguyen, james.nguyen@sas.ac.uk

Abstract: This Element presents a philosophical exploration of the notion of scientific representation. It does so by focusing on an important class of scientific representations, namely scientific models. Models are important in the scientific process because scientists can study a model to discover features of reality. But what does it mean for something to represent something else? This is the question discussed in this Element. The authors begin by disentangling different aspects of the problem of representation and then discuss the dominant accounts in the philosophical literature: the resemblance view and inferentialism. They find them both wanting and submit that their own preferred option, the so-called DEKI account, not only eschews the problems that beset these conceptions, but further provides a comprehensive answer to the question of how scientific representation works. This Element is also available in the Open Access on Cambridge Core.

Keywords: representation, scientific modelling, epistemology of science, model-world relation, DEKI

ISBNs: 9781009009157 (PB), 9781009003575 (OC)
ISSNs: 2517-7273 (online), 2517-7265 (print)

Contents

1 Introduction

1.1 Models

Imagine that you're a shipbuilder working with ocean liners like SS *Monterey* (Figure 1a). Matson Navigation Company has re-purchased the liner from the US government, to whom they had sold the ship following its utilisation during the Second World War. The company now deems the ship too slow for its San Francisco–Los Angeles–Honolulu run, and tasks you to redesign its engine so that it will be able to sail at a certain speed whilst carrying a certain load. What is the minimal power the engine must have to ensure it's up to the task? You could of course just make a guess, install a certain engine, and see whether the ship runs at the right speed when it's back in the water. If you're lucky, the ship works as it is supposed to. But there's a good chance it won't, and that would be a costly failure. A better way to proceed is to construct a model of the ship, a scaled-down version of the real ship you're overhauling, and perform experiments on that model. The model must be carefully constructed: it has to have a shape that reflects the shape of the full-sized ship you're ultimately interested in. And the experiments that you perform on the model have to be carefully designed. In this case, you want to measure the complete resistance, R_C, faced by the model ship as it is propelled through the basin at velocity V, because this gives you information about how powerful the engine needs to be. Experiments of this kind are standard practice in the process of designing ships. In Figure 1b, we see a model ship being moved through a towing tank.

But however carefully you construct your model, and however carefully you perform experiments on it, you're not investigating the model for its own sake. Ultimately, you want to use the results of your investigations to inform you of another system: SS *Monterey* at sea. So you have another task at hand: you have to translate your experimentally discovered facts about the model into claims about the full-sized ship. This translation procedure is subtle and complex: it is informed by our theoretical background knowledge about fluid mechanics, and clever ways of thinking about things like scale, length, and resistance. We'll come back to these later in the Element. For now, what's important is the general pattern of reasoning: the shipbuilder first constructs a scaled-down model of the ship, investigates how the model behaves, and then translates facts about their model into claims about the actual ship.

Now leave ships behind and imagine you've landed a job as a stunt planner for the next instalment of the 007 franchise. You are planning an exhilarating car chase through the streets of London, culminating in the British spy launching

Figure 1a The target system: An ocean liner like SS *Monterey*

Figure 1b Model ship in towing tank

Figure 2a Tower Bridge half open[1]

their Aston Martin across Tower Bridge.[2] The bridge is a drawbridge and the car is supposed to jump over the bridge when it's half open, as displayed in Figure 2a. The producers tell you the angle, α, at which they want the leaves (i.e. the arms) of the bridge for dramatic effect. How fast does 007 have to drive at the jump off point to ensure that the car lands safely on the other side?

Unlike our shipbuilder, you don't produce a scale model of the situation in which you make a small remote-controlled model car jump across a model bridge. Instead, you revert to the power of the imagination and the laws of Newtonian mechanics. You imagine a scenario with two inclined planes facing each other with a gap in the middle. You then imagine a perfect sphere moving up one of the planes with constant velocity v. You imagine that both the planes and the sphere are on the Earth's surface, that all other material objects in the universe have vanished, and that the planes and the sphere are in a vacuum. On the basis of these assumptions, the Earth's gravity is the *only* force acting on the sphere; that is, the sphere is not subject to all the other forces that a real car jumping across a real bridge would experience, such as air resistance and the

---

[1] Credit: 'Tower Bridge Open' by Tony Hisgett from Birmingham, United Kingdom is licensed under CC BY 2.0

[2] Readers might be familiar with similar stunts, including a jump over Tower Bridge in the 1975 movie *Brannigan*; the jump on the 95th Street Bridge in Chicago in the 1980 movie *The Blues Brothers*; the bus jump over an incomplete overpass in the 1994 movie *Speed*; or the jump involving two cars simultaneously in the 2003 movie *2 Fast 2 Furious*.

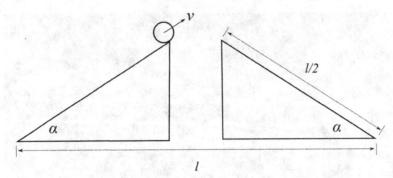

Figure 2b Sketch of the bridge jump model

gravitational pull of other pieces of matter in the universe. This is your model of the bridge jump, where, of course, the two inclined planes stand for the two leaves of the bridge and the sphere for the car. The model is sketched in Figure 2b.

You now apply Newtonian mechanics to your model and find that the equation of motion for the sphere is:

$$m\begin{pmatrix} \ddot{x} \\ \ddot{y} \end{pmatrix} = \begin{pmatrix} 0 \\ -mg \end{pmatrix},$$ (1)

where the x is the horizontal, and y the vertical, coordinate of the ball; g is the gravitational constant on the surface of the Earth; m is the mass of the sphere; and the two dots on x and y indicate the second derivative with respect to time (and recall that this second derivative of a position is acceleration, and so $\ddot{x}$ and $\ddot{y}$ correspond, respectively, to the sphere's acceleration in the horizontal and vertical directions). According to this equation, the sphere moves at a constant horizontal velocity, and accelerates towards the ground at a rate equal to mg. The general solution to this equation is:

$$\begin{pmatrix} x \\ y \end{pmatrix} = \begin{pmatrix} v_x t + x_0 \\ -\dfrac{g}{2}t^2 + v_y t + y_0 \end{pmatrix},$$ (2)

where x_0 and y_0 are the initial conditions (that is, the position of the sphere when it starts the jump); v_x and v_y are the components of the sphere's velocity in the x-direction and the y-direction, respectively; and t is time. From this general solution, you can calculate the minimal velocity that the sphere must have to land on the other inclined plane without falling into the gap between them:

$$v = \sqrt{\frac{gl(1 - \cos \alpha)}{\sin 2\alpha}}.$$ (3)

Given the angle α that the producers want for dramatic effect, and the length l of the bridge, this formula tells you the minimum velocity with which the sphere must move up the inclined plane to fly across the gap and land safely on the other side.

But real cars (even Aston Martins) aren't spherical and don't move in a vacuum. And you're not interested in spheres in vacuums per se. What you really want to know is the velocity at which the actual car has to move up the half open bridge to avoid plummeting into the Thames. So the question is: what can you learn about real cars jumping across half open drawbridges from spheres moving on inclined planes in a vacuum? To answer this question, you have to face the task of translating facts about the model into claims about the actual situation, just like you did when you faced the task of redesigning a ship. Again, we will see in Section 4 how this should be done.

Despite their differences, both of these scenarios rest on a common style of reasoning. You investigate one system, a *model*, and use the results of that investigation to inform yourself about another system, a *target*. In the first case, the model is a concrete physical system itself; it is constructed out of steel (or wood, or paraffin wax, depending on the era) and the shipbuilder performs physical experiments directly on the model. In the second case, the model is a combination of an imagined scenario and mathematical equations (derived from Newtonian mechanics), which the stunt planner can investigate, and there are facts about this model and the solutions of its equations. But neither the shipbuilder nor the stunt planner is interested in their models per se; they are interested in what the models are about. So to reach the end point of their investigations they have to translate their model results into claims about another system: the full-sized ship moving through water and the car jumping the bridge.

The aforementioned examples are not isolated instances of some outlandish style of reasoning. Models are used across the sciences; they are one of the primary ways in which we come to learn about the world. Scientists construct models of atoms, black holes, molecules, polymers, populations, DNA, rational decisions, financial markets, climate change, and pandemics. Models provide us with insight into how selected parts or aspects of the world work, and they act as guides to action. Much of scientific knowledge, and understanding, is ultimately based on the results of some modelling endeavours.

How do models work? How can the investigation of a model possibly tell us anything about something beyond the model, some system out there in the world? Our answer here is that models do this because they *represent* their targets. Just as the scale model in the shipbuilder's tank represents the full-sized ship and the stunt planner's model represents the actual car, the physicist's model of a black hole represents parts of the universe from which no light can escape; the economist's macroeconomic model represents the actual economy of a given country; and the epidemiologist's model represents how a disease will spread through a country as a result of various policy interventions by a government.

As these examples suggest, models can be used to represent *particular* target systems like a particular government's economy, a specific ship, and so on. But they can also be used to represent *types* of targets. Depending on the details, economists might employ models to reason about economies in general; a physicist can use a model to represent a type of atom like hydrogen; the model ship can be used to represent a certain type of ship; and the mathematical model of 007's stunt can also be used to represent a type of stunt involving car jumps across bridges (a few exemplars of which are mentioned in Footnote 2). So by 'target system' we can mean both specific systems and types of systems.

In each of these cases the model stands in for 'the' target system; the model is the secondary system that scientists investigate, with the hope that the results of their model-based investigations will deliver insight into their targets. So in order to understand how model-based reasoning works, we need to understand how models represent. In this Element, we provide a philosophical investigation into this question.

At this point, one might worry that our focus on models is too narrow. Scientists use plenty of other kinds of representations to reason about systems in the world: doctors use MRI scans to learn about brain structure; particle physicists pore over bubble chamber photographs to learn about the nature of subatomic particles; and astronomers study the images produced with telescopes. But it pushes the limits of language to deem any of these kinds of representations models. Moreover, there are plenty of non-scientific representations that function in a similar way. All of us are familiar with using maps to navigate new cities, and we regularly make inferences about the subjects of photographs based on features of the photograph. In general, we call a representation that affords information about its target an *epistemic representation*, and it is clear that models are just one kind of such representation. Given this, there is a question whether our analysis in this Element covers things beyond models; whether it covers epistemic representations more generally. By and large we think it does, and most of the existing accounts of how models represent end up being accounts of epistemic representation more generally.[3] In this Element, we primarily focus on models because models are crucial to the

---

[3] See the discussion of the representational demarcation problem in Frigg and Nguyen (2020, Ch. 1), and the discussion of how the different accounts handle demarcation in later chapters of that book.

working of modern science, and because discussing different kinds of representations side by side would end up using more space than we have. We will briefly broaden our scope in Section 4, and readers who are interested in other kinds of epistemic representations – images, and certain works of art, for example – are encouraged to consider how what we say applies to these as they proceed through the following sections.

1.2 Questions Concerning Scientific Representation

At first glance it might seem like there is only one question to be asked here: how does a model represent its target? But looking a little closer, we see that this question breaks up into several different questions. To have a clear focus in our investigation, it's important to disentangle these and clarify how answers constrain the shape of the conceptual landscape concerning how models work. This is the project for this section, in which we lay out the relevant questions; in the next, we discuss what it takes to answer them appropriately.

The first, and most fundamental, question to investigate is: in virtue of what does a model represent its target? Call this the *Semantic Question*.[4] Answering this will allow us to understand how the steel vessel that is dragged through the towing tank comes to represent a real ship, and how the shipbuilder manages to translate results of model experiments into claims about a full-sized ship; likewise, it will allow us to understand how an imagined scenario consisting of two inclined planes and a sphere in a vacuum comes to represent a real-world bridge jump, and how the stunt planner manages to plan the jump based on the model.

It's of paramount importance that we don't confuse this question with a closely related one, namely: what makes a model an *accurate* representation? A model can represent its target, without doing so accurately. To see this, alter the aforementioned examples slightly. You may not have a very good initial idea of the ship's shape because you haven't been able to measure it up and no plans are available. So you may decide to use an empty barrel as a model of the ship. When you finally see the ship, you realise that this is a bad model because it doesn't have the form of the ship at all. Nevertheless, the barrel is a representation of the ship; it's just not an accurate representation. Or assume that an error has been made in measuring the angle of the open leaves of the drawbridge, and the angle in reality is twice the angle in your model. As a result, the model will underestimate the velocity needed to get to the other side, and the

---

[4] We're using the term 'semantic' in the broad sense of referring to the relationship between symbols and what they are about. Thus understood, a discussion of how models relate to their targets falls within the scope of semantics.

car, along with the stunt driver, will plummet into the river. If this happens, the model is not an accurate representation of the target system, but it is a representation of it nevertheless.

The lesson is that we should distinguish between the question of what turns something into a representation of something else to begin with (the Semantic Question), and the question of what turns something into an accurate representation of something else. Call the latter the *Accuracy Question*. The Semantic Question is conceptually prior in that asking what makes a model an accurate representation presupposes that it is a representation in the first place: a model cannot be a *mis*representation unless it is a representation. But once this has been established, there is a genuine question about what it takes for a representation to be accurate.

The distinction between representation and accurate representation is not an artefact of the simple examples we have used to illustrate it. The history of science provides us with a wealth of examples of inaccurate, but nevertheless representational, models. Ptolomy's model of the solar system represents the solar system even though it is inaccurate with respect to the orbit of the Earth around the Sun. Thompson's 'plum pudding' model represents atomic structure, but it is inaccurate with respect to the distribution of charge within an atom. Fibonacci's model of population growth is inaccurate with respect to the long-term growth of a population because it assumes that organisms are immortal and food supplies are unlimited. But despite their inaccuracies, all of these models represent their targets.

So far in this section we've distinguished between the question of what makes a model represent, and what makes it accurate. But some might worry that there's an even more prior question lurking in the background: what is a 'model'? Call this the *Model Question*. In the case of the model ship the answer is relatively straightforward: the model is the concrete material object towed through the tank. But many models aren't like this; they are, to use Ian Hacking's memorable phrase, things that we 'hold in our heads rather than our hands' (1983, 216). It has become customary to refer to such models as 'non-concrete' models. The model of the car jump is of this kind. Earlier we said it was an imagined scenario combined with mathematical equations from Newtonian mechanics, and we also said that there were facts about this model that the planner would investigate. One option then, would be to identify this model, and others like it, with mathematical entities (and then leave mathematicians and philosophers of mathematics to work out what they are). But there's a lingering worry that this isn't the whole story. The car jump model might involve mathematics, but it's not obviously purely mathematical. Before writing down equations, the stunt planner had to imagine a scenario with a sphere moving in the vacuum on an inclined plane. And it's not clear that this can be accounted for by identifying the model with something purely mathematical.

A philosophical account of modelling should have something to say about how we should think about models of this kind. This goes beyond metaphysical bookkeeping; as we will see, how one answers the Model Question has implications for how we understand the Semantic Question and the Accuracy Question.

1.3 What Does Success Look Like?

The next thing to establish is the success conditions on answers to these questions. What counts as a successful answer to the aforementioned questions?

We begin with conditions on answers to the Semantic Question. First and most straightforwardly, there is a *direction* to the representation relationship that holds between models and their targets. Typically at least, models represent their targets, but not *vice versa*: the scale model in the towing tank represents SS *Monterey*, but SS *Monterey* doesn't represent the scale model. We say 'typically' here because we are not assuming that it's a conceptual impossibility; in some special cases a representational relationship can hold both ways. Rather, we require that any answer to the Semantic Question should not entail that the model–target representation relation is always symmetric. We call this the *Directionality Condition*.

Another important condition of success on answering the Semantic Question is accounting for the fact that models are *informative* about their targets. Some representational relationships do not work this way: the term 'atom' can be seen as representing (at least in some sense) atoms, but it's uninformative: no investigation into the term itself will allow us to extract any information about what it refers to. In contrast, a model represents its target in a way that does allow such information extraction (although, of course, that information doesn't have to be true, because models don't have to be accurate). To use Swoyer's (1991) phrase, models allow for *surrogative reasoning*: by investigating the behaviour and features of the model, scientists can generate claims about the behaviour and features of its target. We call this the *Surrogative Reasoning Condition*.

The distinction between representation and accurate representation motivates a further condition for success: any answer to the Semantic Question should be compatible with the fact that models can misrepresent their targets; no answer should entail that all representations are accurate, nor that inaccurate models are non-representations. In brief, a viable answer to the Semantic Question must distinguish between misrepresentation and non-representation. This is not to say that all models are inaccurate (although there is reason to think that no model represents with perfect accuracy). Indeed, much of the motivation for investigating how models work stems from the fact that at least some of them are accurate; some of them are paradigmatic instances of the cognitive success of the scientific endeavour. But it remains that some models misrepresent, and thus whatever it is that establishes a representational relationship should not equate representation with accurate representation. We call this the *Misrepresentation Condition*.

The previous condition relates to models that represent actual targets in the world (whether specific systems, or types of systems), but do so inaccurately. We should also recognise that some models don't represent any actual target whatsoever. Straightforward examples of models of this sort include engineering models of structures never built. Vary our initial example slightly and assume that you are tasked with designing a new ship rather than redesigning an existing one. But for some reason the ship is never built. In that case your model doesn't represent anything in the world. The same goes for architectural models of buildings that have never been constructed and models of spacecraft that have never been realised. Targetless models aren't unique to fields that are in the business of constructing something; such models also appear in theoretical science. Population biologists construct and investigate models involving a population consisting of four different sexes to see how such a population would develop; elementary particle physicists study models of particles that don't exist to learn about techniques like renormalisation; and philosophers of physics construct models in accord with the principles of Newtonian mechanics in order to demonstrate that the theory is consistent, under certain conditions, with indeterminism, without the model representing, or indeed being intended to represent, any system in the world.[5] A philosophical account of modelling should accommodate models that don't have real-world targets. We call these 'targetless models', and the condition that they be allowed for the *Targetless Models Condition*.

We now turn to the Accuracy Question. Unlike truth, which (many people think) is an all-or-nothing matter, accuracy comes in degrees. Models can be more or less accurate, depending both on the scope of the features they represent, and on how well they represent those features. And the fact that a model misrepresents some aspects of its target doesn't entail that it misrepresents all aspects. For example, whilst the Ptolemaic model of the solar system might misrepresent the structure of the celestial orbits, it accurately represents, to some degree, the relative position of the celestial bodies as seen in the night sky

---

[5] For discussions of *n*-sex populations, see Weisberg (2013). For discussion of elementary particles, see Hartmann (1995). For discussions of Newtonian mechanics and indeterminism, see Norton (2008) for 'Norton's Dome' and Xia (1992) for a model involving charmingly named 'space-invaders'.

from the surface of the Earth. So its accuracy falls somewhere between perfect accuracy and total misrepresentation. Any tenable notion of accuracy must allow for, and indeed make sense of, such gradations. We call this the *Gradation Condition*. This condition comes into contact with the Misrepresentation Condition discussed earlier. The less accurate a model, the more it misrepresents. So requiring that any answer to the Semantic Question allow for misrepresentation entails requiring that any answer to the Accuracy Question allow for there being models with low-grade accuracy, and *vice versa*.

What counts as accurate depends on the context in which a model is used, and on the aims and purposes of the model user. An architectural model that provides measurements that are correct within a ±5 mm error margin is extremely accurate; a molecular model that predicts the extension of a molecule with a precision of ±5 mm is an egregious failure. Indeed, the very same model may count as accurate for one purpose, but not for another. This happens, for example, when context restricts which features of a target system must be represented accurately in order for the model to count as accurate. As Kuhn notes, in the context of navigation and surveying, the Ptolemaic model is still employed today, and the model is accurate for these purposes (Kuhn 1957, 38). But in the context of using the same model to explain why planets appear where they do, the model is inaccurate. In general, an assessment of what counts as accurate must take contextual factors into account. We call this the *Contextuality Condition*.

The Model Question comes with its own associated conditions of success. Recall what it asks: what are models? Any answer to this question has to help us understand what scientists are talking about when they're talking about their models, especially when they are talking about the model as having an, in some sense, independent existence from any target system it represents.

There is right and wrong in the model: certain claims are true in the model and some are false. A first condition on a successful answer to this question therefore is that it provide an account of *truth in a model*, *model-truth* for short, and accordingly we call claims that are false in a model *model-false*. In certain cases, model-truth corresponds to what we standardly mean by 'truth'. To say that it is model-true that the concrete ship model experiences a complete resistance of a certain strength R_C is the same as to say that it is true that it experiences that resistance. In general though, model-truth is not the same as truth simpliciter. The claim 'the sphere moves on an inclined plane' is model-true, but false in the world (there is no such sphere!). Moreover, it is model-false that it loses mass on the way. But on what basis are some propositions model-true, and others model-false? As we have seen previously, there is no sphere

moving up an inclined plane, and so the claim cannot be true in the same sense
in which 'the car moves on Tower Bridge' is true.

This question is pressing for two independent reasons. First, we often attribute *material* features, like having a certain velocity or being acted on by
gravity, to non-concrete models. But if these models aren't actually existing
physical objects, then in what sense, if any, can they have such features? Notice
that this problem remains even if one answers the Model Question by identifying such models with mathematical systems: even though mathematical objects
exist in some sense, they're not obviously the sort of things that can have
material features (gravity doesn't act on mathematical objects!). Second, let
us call the description that is used to introduce the model the *model-description*.
Sometimes we can settle the question of whether a proposition is true in the
model by appeal to the model-description: if the proposition is contained in the
model-description it is model-true. But we often care about model-truths that
the model-description remains silent about. We might say that 'the sphere
moves on an inclined plane' is model-true because the model-description says
that it does. But the model-description does not contain anything like 'the
sphere moves in a parabolic orbit once it leaves the inclined plane', and so
this route is foreclosed in this case. Since we are mainly interested in claims that
are not explicitly stated in the original model-description (recall that the main
result that the model delivers is Equation (3), which is not stated in the model-
description!), an answer to the Model Question must help us understand how
claims made about models, including those attributing material features to
them, can be model-true (or model-false). Call this the *Model-Truth Condition*.

Modellers spend a lot of time investigating model-truths. Models are tools of
investigation and they can serve this purpose only if the users of models can
come to know what is true in the model and what isn't. If there is no way for you
to find out what resistance the model ship experiences when it's dragged
through the tank, or how far the sphere will fly once it's airborne, then the
models are useless. For this reason, the next condition concerns the *epistemology of models*. Any story concerning how to understand model-truth must be
accompanied by a story about how we can come to know these model-truths and
how we justify our findings. Call this the *Model-Epistemology Condition*.

Third, an answer to the Model Question should provide *identity conditions*
for models: under what conditions are scientists talking about, and investigating, the same model? The ship model can be made from metal or from wax. So
they would be different material objects, but would they still be the same
model? Problems concerning identity also arise with non-concrete models.
It's commonplace that one can describe the same object in many different
ways, and this also goes for models. Someone else could have described the

car jump model using a different set of sentences than the ones used at the beginning of this section, and yet they could have described the same model. But on what grounds do we assert that this alternative model-description would really describe the same model? In the case of non-concrete models, we can't clear up potential ambiguities by simply pointing to an object and say 'that's what I'm talking about'. Non-concrete models are given to us only through model-descriptions, and so there is a question about when two model-descriptions specify the same non-concrete model. Any answer to the Model Question must provide the conditions under which models are identical. Call this the *Model-Identity Condition*.

The questions and conditions that we have introduced so far, along with the relations between them, are summarised visually in Figure 3. We think that these are some of the most important questions and conditions, but we do not algorithmically consider them in order for each account in what follows. It is recognised that competing accounts of scientific representation engage with these questions and conditions in different ways, and as such we follow the existing literature in emphasising which of them are particularly pertinent for the accounts in question. Moreover, we don't claim that this list is exhaustive. Indeed, in Frigg and Nguyen (2020) we discuss a number of additional issues that concern the use of mathematics in the empirical sciences, the use of different representational styles, and the relation between scientific

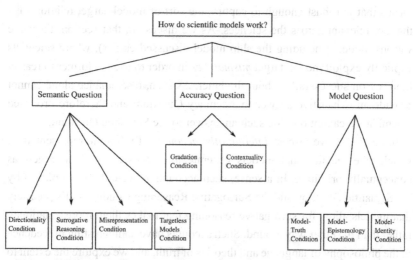

Figure 3 The questions and conditions an account of representation must answer. The simple lines indicate that the question at the top breaks up into three different questions; the arrows indicate that a viable answer to the question must meet the condition that the arrow points to.

representation and other kinds of representation, including works of art. A discussion of these issues presupposes answers to the questions we have introduced in this section, and so dealing with the issues we have introduced here provides the groundwork for further discussions.

1.4 Roadmap

In the coming sections we investigate different accounts of scientific representation discussed in the philosophical literature.

In Section 2, we discuss *resemblance accounts* of scientific representation. These accounts are based on the idea that models, in some sense, resemble, or are supposed to resemble, their targets. We use 'resemblance' as an umbrella term for any kind of likeness between a model and its target. Some accounts focus on a kind of resemblance that is based on material features, the kind we invoke when we say, for instance, that London buses are similar to phone boxes because they are both red. We call resemblance of this kind *similarity*. An alternative version of resemblance understands model–target resemblances in terms of *structural* relationships, which are specified by different mappings, such as isomorphisms, between the two.

As we will see, such accounts of representation are most plausible when the resemblances in question are supposed to answer the Accuracy Question, with *proposed* resemblances answering the Semantic Question. However, a major question remains whether there is a way of explicating what 'resemblance' means that is robust enough to capture the sort of model–target relationships that are relevant across the sciences. As we discuss in that section, there are various models (including the ship model discussed earlier), where scientists explicitly exploit model–target *mismatches* in order to use the former to reason (and reason successfully) about the latter. As such, resemblance alone cannot provide an exhaustive answer to Accuracy Question, and therefore proposed resemblance cannot provide such an answer to the Semantic Question.

In Section 3, we discuss *inferentialist accounts* of scientific representation, which take the fact that models generate inferences about their targets as conceptually primitive. In a sense, such accounts reverse the roles played by the Semantic Question and the Surrogative Reasoning Condition: it's precisely that models allow for surrogative reasoning that makes them representational, rather than the other way around. Such accounts have an interesting precedence in the philosophy of language and theories of truth, and we explore the extent to which they can be successfully deployed for understanding how models represent. We ultimately conclude that lessons from other areas of philosophy fail to motivate the idea that the Semantic Question can be put aside in its entirety, and that inferentialist accounts leave important questions open.

In Section 4, we present our own preferred account of scientific representation. We introduce the account via a close analysis of how you would reason were you to use the ship model to generate information about SS *Monterey*: you use the model to *denote* the target ship; the model *exemplifies* various features; you deploy a *key*, which turns exemplified features into features that are reasonable to *impute* to the target ship. This provides the 'DEKI' account of scientific representation. We explicate what each of these conditions mean, and we demonstrate how they can be combined to yield a complete account of scientific representation that answers each of the questions presented in this section in a way that meets the associated conditions. We then turn to the bridge jump model as another illustrative example, this time one involving a non-concrete model, paying attention to how the details of the conditions (the key in particular) are realised in that case. This demonstrates how the account is 'skeletal' in the sense that its conditions need to be filled in on a case-by-case basis in order to play an informative role. In providing examples of how this is done, we hope to show the practical value of thinking about epistemic representation through the lens we provide. Our account provides more than conceptual bookkeeping; it offers normative lessons for the 'best practices' in scientific modelling, and it suggests refocusing at least some of our philosophical investigations into how models work. To understand a model, we need to go beyond the model itself; we need to understand the disciplinary contexts and practices in which scientists reason with the model, and the rules that they implicitly subscribe to in doing so.

2 Resemblance and Representation

2.1 Introduction

It's easy to be led to the idea that representation has something to do with resemblance: what makes a photograph a representation of its subject if not the fact that the two look alike? And philosophical accounts of representation that invoke resemblance have a long history, going back to Plato's *The Republic* (Book X).

Scientific representation also appears to have something to do with resemblance. Consider the examples in the previous section. The scale model of the ship resembles the full-sized ship in some respects. After all, when you designed the model, you ensured that it had the same geometric shape as its target. Moreover, this resemblance isn't accidental to the way in which the model works: you relied on it when you drew inferences from the behaviour of the model to the behaviour of the target. If the model were a different shape, then it would be unlikely that you would be able to extract any useful information about the full-sized ship from the model.

The same considerations apply to the model of the car jump. Here the model–target resemblance doesn't concern what the two look like. But nevertheless, the model has a certain structure that, in some sense at least, seems to resemble the structure of the target. Both involve an object moving on an inclined surface, with a certain trajectory determined by the influence of gravity and Newtonian laws of motion. And when you reason with the model as a stunt coordinator, you work hard to ensure that the model trajectory matches the trajectory of the car. Without this, you threaten to put the lives of the stunt performers at serious risk.

These sorts of considerations suggest that when a model user draws surrogative inferences about the model's target, they exploit the fact that the model and the target resemble one another. And given the tight connection between surrogative reasoning and scientific representation, this would suggest that these resemblances have something to do with how scientific models represent. The question, then, is where in the conceptual landscape introduced in the previous section should we locate resemblance? Does it answer the Semantic Question or the Accuracy question? And what sort of constraints does it place on answers to the Model Question?

In order to start addressing these questions, we first need to establish what we mean by 'resemblance', which, as it turns out, is no straightforward task (Section 2.2). Once we've explicated some options, we can then turn to investigate how resemblance might be put to work in helping us understand scientific representation (Section 2.3). The most viable suggestion is that resemblance relationships be invoked to answer the Accuracy Question: a model is accurate to the extent that it resembles its target (Section 2.4). Finally, we turn to the Model Question and see what restrictions the resemblance view imposes on the ontology of models, and indeed target systems (Section 2.5).

2.2 Resemblance

In order to understand how resemblance relates to representation, we first have to clarify what it takes for two systems to 'resemble' one another. And this is more than philosophical bookkeeping: as we will see, there are serious issues with relying on an intuitive understanding of the concept. In general, two things resembling each other means that they share some features. The model ship resembles the full-sized ship in the sense that they have the same shape; a London bus resembles a London phone box in the sense that they are of the same colour; and the copy of *The Sunday Times* on the front doorstep resembles all the other copies that day in the sense that they are made out of the same material and exhibit the same distribution of ink on paper.

For our current purposes it is useful to distinguish between two different kinds of features that can be relevant to resemblance. First, we can consider the sorts of homely *material* features we have in mind when we say that two objects are *similar* to one another. These features include visual features like colour, shape, and distribution of ink on paper, but in terms of scientific models they also include more general material features like having a certain volume, being subjected to a certain resistance, being acted on by gravity in a certain way, and so on. Second, we can focus on 'structural' features, which, broadly speaking, concern the arrangement of features rather than the features themselves.

We return to material features later in the text. At this point it is important to say more about what is meant by structural features. In the context of thinking about model-based science, there is a rich tradition in thinking about models as *structures*. These are mathematical entities, consisting of a *domain* made up of a set of elements and an ordered set of *relations* defined over them.[6] Thinking about model–target relations in these terms amounts to thinking about resemblance with respect to their structural features. For our current purposes, what's important about this notion of structure is that the intrinsic nature of the elements of the domain, and the intrinsic nature of the relations defined upon them, are of no consequence. Consider, for example, a system consisting of a grandfather, his daughter, and her son. The structure whose domain is those three people, with the relation *is a descendant of* is the exact same structure as the structure with the same domain and with the relation *is younger than*. And if we consider three books of different lengths and the relation *has fewer pages than*, the books have the same structure as our family. Even though the 'meaning' or 'intension' of the relations is different in each case, they are the same from an abstract point of view: there are three objects with a relation on them, and that relation is irreflexive, asymmetric, and transitive. So we can say that a structure consists of dummy objects (because it only matters that there is something and it doesn't matter whether the something are people or books or anything else) with purely extensionally defined relations on them (because it only matters between which objects the relation holds and not whether the relation is *is a descendant of*, *is younger than*, or *has fewer pages than*).[7]

---

[6] We use 'relation' here to include both one-place relations, that is, properties, and functions. In the case of the bridge jump model, such a structure consists of the real numbers, with Equation (2) acting to specify a trajectory, thought of as a relation (between a 'time' coordinate to a pair of x and y coordinates) on them.

[7] For a clear discussion of this aspect of structures, see Russell (1919/1993, Ch. 6).

If we have two structures, we can ask whether they resemble one another with respect to their structural features. One way to make this precise is to ask whether there exists a function from the domain of one of the structures to the domain of the other (i.e. a way of associating each element of the former with one element of the latter) that preserves the relations defined on the former. One particularly important kind of such function is an *isomorphism*, which requires complete agreement with respect to structural features. To illustrate this, consider the four structures symbolically represented in Figure 4.

S_1 consists of a domain of three objects (symbolised by the dots), with the top object having a property (symbolised by the black circle around it), and the bottom and middle objects being related by a two-place relation (symbolised by the line with the arrow running from one to the other). S_2, S_3, and S_4 are also structures, with different domains and properties and relations defined on them (where objects, properties, and relations are symbolised in the same way as in S_1).

We can see that S_1 and S_2 share their structural features in the sense that they are isomorphic: they both consist of three objects, a property, and a two-place relation, and these properties and relations are distributed across the domains in

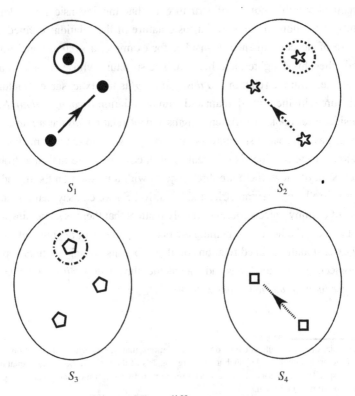

Figure 4 Four different structures

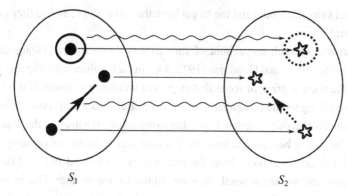

Figure 5 Isomorphic structures

the same way. This can be made precise by considering the arrows from S_1 to S_2 in Figure 5. The straight arrows associate each element of the domain of S_1 with a unique element of the domain of S_2 in such a way that no two objects in S_1 map to the same object in S_2, and all of the objects in S_2 have a corresponding element in S_1. This means that the arrows specify a mapping that is a *bijection*. The wiggly arrows symbolise how this function preserves the structure of the property and relation in S_1. The object with the property is mapped to an object with the corresponding property (the dotted circle in S_2), and the objects related by the two-place relation are mapped to objects also related by a two-place relation (the dotted line with the arrow in the middle in S_2). The fact that the function is a bijection that preserves the property and relation means that it is an isomorphism. So, S_1 and S_2 are isomorphic to each other; they share the exact same structural features.

It is also clear that there is no such isomorphism between S_1 and either S_3 or S_4. If we compare S_1 to S_3, there is no corresponding two-place relation in the latter and so no function from S_1 to S_3 can preserve that relation. And if we compare S_1 to S_4, we can see that any function from the former to the latter will have to be such that at least two elements map to the same element, and so there is no bijection between the two domains.[8]

Now let's return to the first notion of resemblance we discussed, resemblance with respect to non-structural, that is, material, features. This sort of resemblance is what we call 'similarity'. In the first instance, we might then say that two objects are similar if and only if they share at least one feature. In the case of

---

[8] Isomorphism is the strongest kind of structural resemblance, and there are weaker mappings that have been invoked in this context (homomorphism, partial homomorphism, partial isomorphism, isomorphic embedding, etc.). For our current purposes, we set them aside; for definitions, see Hodges (1997) and French and Ladyman (1999); and for critical discussions, see Pero and Suárez (2016) and Pincock (2012).

the model ship, the model and the target have the same shape, and so they count as similar.

Unfortunately, such an account of similarity is a non-starter. As argued by Quine (1969, 117–8) and Goodman (1972, 443) in their influential critiques of the notion, the mere sharing of some feature makes similarity vacuous: if this were right, everything would be similar to everything else. To see why, consider some pair of objects that don't seem to be similar to one another: a mug on the desk and the dog in the dog bed, for example. With a little ingenuity we can come up with a material feature that they share, for example, being located in East London. Thus, according to the proposal, they are similar to one another. This is not an artefact of the example: for any pair of objects it is not too difficult to specify some material feature that they share, even if we have to resort to things like being earthbound, being located in space-time, being thought about by you when you are reading this Element, and so on. Given that it's too easy for any pair of objects to find a material feature they share, it's too easy to conclude that they are similar.

One thing that has gone wrong here is that there is no restriction on which shared material features can be considered to establish the similarity relation. Rather than requiring that two objects share arbitrary features to be considered similar, we might require that they share *relevant* features, where what counts as 'relevant' depends on the context in which the similarity is being considered. If location isn't relevant in a particular context, then the mug and the dog won't count as similar (in that context) on grounds that they are located in East London, and likewise for the other examples.

In the context of accounts of scientific representation, this seems to be the notion of similarity that Giere has in mind when he states 'since anything is similar to anything else in some respects and to some degree, claims of similarity are vacuous without at least an implicit specification of relevant *respects* and *degrees*' (Giere 1988, 81).[9] By appealing to the role that context plays in specifying the features that are relevant to similarity, the issue of spurious similarities is avoided.

However, this account faces another problem concerning the distinction between two objects being similar in the sense that they share the exact same relevant feature and being similar in the sense that they each instantiate a feature, where the features themselves are 'similar'.[10] To illustrate the latter, consider again the similarity between a London bus and a London phone box,

---

[9] For now we're focusing on the 'relevant respects'. For more on relevant degrees, see later in the text, and also Weisberg (2012, 2013, Ch. 8), who invokes Tversky (1977) to provide an explication of this version of similarity.

[10] Niiniluoto (1988, 272–4) calls the former 'likeness' and the latter 'partial identity'. A similar distinction appears in Hesse's work on analogies (1963, 66–7).

where the relevant similarity concerns their colour. If we look closer, we'll see that the bus and the phone box aren't the exact same colour: the phone box is a shade darker than the bus. Hence, there's no relevant colour property that they share, which is what would be required by the idea under consideration. Accordingly, it looks like we're forced to conclude that the two aren't similar with respect to their colour after all, which seems wrong. The similarity we are interested in here is located at the level of properties themselves: the red of the phone box is similar to red of the bus. But now we wonder whether, and if so how, this kind of similarity can be analysed in terms of two objects instantiating the same property.

Parker (2015), in her discussion of Weisberg's account of scientific representation, provides another relevant example. Here the relevant similarity concerns whether the US Army Corps of Engineers' physical scale model of the San Francisco bay area and the bay area itself are similar with respect to relevant hydrodynamical parameters such as their Froude number (a dimensionless parameter quantifying flow inertia). Due to various issues surrounding how dimensionless parameters are subject to scaling effects, the Froude number of the model and the Froude number of the actual bay aren't exactly the same. But in certain modelling contexts, including the San Francisco bay case, the numbers count as similar as long as the model's parameter is 'close enough' to the target's parameter. The question then is whether shared features accounts can accommodate this kind of similarity.

Parker suggests that one way to make good on the fact that sharing similar, rather than the same, features still counts as a relevant similarity is to turn the former into the latter by introducing the idea of an *imprecise* feature (*ibid.*).[11] The rough idea is that if two objects instantiate similar but non-identical, features, then we can consider a more general feature which both of the features are instances of. When it comes to quantitative features (i.e. features that take numerical values) like the two Froude numbers, this can be done via introducing interval-valued features of the form: 'the value of the feature lies in the interval $[p-\epsilon, p+\epsilon]$' (*ibid.* p. 47), where p is the value of the parameter in the target, and ϵ specifies how precise the overlap needs to be.[12]

Whilst this might work for features of this kind, it is unclear whether this approach can be applied to all instances of similarity at the level of features themselves. The aforementioned approach won't work in the context of

---

[11] An alternative solution would be to adopt a so-called *geometric* account of similarity. For a general discussion, see Decock and Douven (2011).

[12] Weisberg (2015) accepts that this is what is required by his account, since it is an attempt to reduce the notion of similarity between features to that of shared features. For more on how these problems relate to this sort of reduction, see Teller (2001) and Khosrowi (2020).

qualitative features (i.e. features that don't take numerical values), which require some other way of introducing imprecision. It's not obvious what this would be (consider, e.g. the difficulty in assigning a number measuring how 'rational' an agent is, and using this number to measure the 'similarity' between a model-agent reasoning according to rational choice theory and an imperfect human agent).

Since none of the problems for the resemblance account of representation will depend on how this issue is resolved, let's simply assume that this can be done in some way or another.

We hope it is now clear what we mean by 'resemblance'. Where the resemblance concerns material features, we use the term 'similarity', and when it concerns the structure of the model and the target, we use the term 'isomorphism'. With respect to the former, what's important is that the context provides some restriction on which material features are considered relevant. Moreover, it requires careful treatment in cases where the similarity relationships in question are underpinned by the sharing of similar features, rather than co-instantiating the exact same feature. With respect to the latter, what's important is that the comparison concerns the existence of structure-preserving mappings between models and their targets. With this in mind, we can now turn to how resemblance, thought of in either of these two ways, can be used to understand how scientific models represent.

2.3 The Semantic Question

A straightforward attempt to answer the Semantic Question by invoking the notion of resemblance is the following:

> RESEMBLANCE: A model represents its target if and only if the two resemble each other.

Further to the previous section it is understood that resemblance can be either similarity (in relevant respects) or isomorphism. Such an answer has the benefit of meeting the Surrogative Reasoning Condition: if a model and its target are similar in the sense of sharing relevant features, then, from the fact that a model has a relevant feature, a model user can infer that the target does as well; and if a model and its target are isomorphic, then if the model has a certain structure, a model user can infer that the target does too.

Unfortunately, RESEMBLANCE fails to meet the other conditions, and has a number of other significant flaws. With respect to the Directionality Condition, it has been pointed out that resemblance has the wrong logical properties to establish representation: similarity and isomorphism are both reflexive (everything is similar to itself) and symmetric (if x is similar to y, then y is similar to x), but representation is not.[13] Due to symmetry, all targets also represent their models, which is not the case; and due to reflexivity, all representations are self-representations, which is wrong. Even if self-representation may occur in certain circumstances (Magritte's *The Treachery of Images* provides a nice example), this is the exception rather than the rule. So the logical properties of similarity and isomorphism pose a problem for anyone appealing to resemblance to establish representation.

RESEMBLANCE fails to accommodate the Targetless Models Condition: if a model doesn't have a target, then there is nothing it is similar to. So the proposal simply remains silent about how these models work. Defenders of a resemblance view would then have to provide some further analysis of how such models represent, if they do at all.

The next problem has gained attention due to Putnam (1981, 1–3).[14] In a thought experiment he invites us to imagine an ant tracing a line in the sand that just so happens to resemble Churchill. Putnam asks: does the trace *represent* Churchill? According to him, the answer is 'no'. The ant has never seen Churchill, indeed has no connection to him whatever, and certainly didn't have the intention to represent him. Putnam concludes that this shows that '[s]imilarity ... to the features of Winston Churchill is not sufficient to make something represent or refer to Churchill' (*ibid.*, 1).[15] What is true of the trace and Churchill is true of every other pair of items that resemble one another: resemblance on its own does not establish representation.

A final problem is even more damning. The Misrepresentation Condition requires that models can represent their targets, but do so inaccurately.[16] But if models have to resemble their targets in order to represent them, either via sharing relevant features or being related by an isomorphism, then this precludes the possibility of misrepresentation. If a model has to share some feature with its target to represent it as having that feature, then it cannot be mistaken about whether or not the latter has it (of course, the model could be accurate with respect to another feature, but that doesn't help RESEMBLANCE explain how the model misrepresents features it purports to represent). Alternatively, if a model has to share the relevant structure with a target in order to represent it (which is

---

[13] See Goodman (1976, 4–5). For further discussions, see Frigg (2006) and Suárez (2003). Weisberg (2013) discusses a notion of similarity that is not symmetric; it is, however, still reflexive.

[14] Black (1973) and Suárez (2003) also discuss similar thought experiments.

[15] The same clearly applies to structural relations.

[16] Misrepresentation should be distinguished from idealization, and resemblance views are implicitly committed to the idea that idealization is misrepresentation, since an idealization is a non-resemblance. For further discussion, see Nguyen (2020).

required by the existence of an isomorphism), then it cannot be mistaken about whether the target has that structure.[17] Thus, RESEMBLANCE has difficulty meeting the Misrepresentation Condition: representation is conflated with accurate representation, and cases of misrepresentation are misclassified as non-representation.

For these reasons, RESEMBLANCE is a non-starter. However, the fact that resemblance allows for successful surrogative reasoning in cases where models are accurate representations suggests that it may feature somewhere else in an account of representation. The idea is the following: the intentional act of a model user *proposing* that a model resemblances its target can be deployed to answer the Semantic Question, and then whether or not the two are in fact similar (the proposal can be true or false) can be used to answer the Accuracy Question. This delivers the following response to the Semantic Question:

> PROPOSED RESEMBLANCE: a model represents its target if and only if a model user proposes that the two resemble each other.

Again, it is understood that resemblance can be either similarity or isomorphism. Such an answer seems plausible, and we'll address how it relates to accuracy in the next section, but for now a few comments on how it fares as an answer to the Semantic Question are in order.

A model user's proposal that a model resembles its target is what Giere calls a 'theoretical hypothesis', a statement of the form 'the model and the target resemble each other in these respects (structural or otherwise)' (Giere 1988, 81). These hypotheses may also depend on the purposes for which the model is being deployed (i.e. model users may deploy different hypotheses in different contexts), and on the intentions of the model user. For example, Giere further develops this account and states that we should analyse how models represent using the following schema: '*S* uses *X* to represent *W* for purposes *P*' (2004, 743), or in more detail: 'Agents (1) intend; (2) to use model, M; (3) to represent a part of the world W; (4) for purposes, P. So agents specify which similarities

---

[17] It is worth noting here that the weaker mappings mentioned in footnote 8 fare better in this respect. Two structures may be related by such a mapping without being isomorphic, and thus if a weaker mapping is sufficient for representation, then a model may represent its target (in virtue of the existence of such a mapping), without accurately representing all of its structure (which would be precluded by the lack of an isomorphism). Whilst more promising than an account built on isomorphisms, such an approach still faces difficulty accounting for how models can misrepresent *particular* structural features of their targets: if, in order for a model to represent some particular structural target feature, there must be a mapping from the model to the target that preserves that feature, then it remains unclear how such a feature could be represented inaccurately. See Frigg and Nguyen (2020, Sec. 4.4) for further discussion, and Pincock (2005) and French (2021) for useful discussions about whether and how the partial structures approach can account for these difficulties.

are intended and for what purpose' (Giere 2010, 274). Other authors have presented similar views. Van Fraassen offers the following as the 'Hauptstatz' of a theory of representation: '*There is no representation except in the sense that some things are used, made, or taken, to represent things as thus and so*' (2008, 23, original emphasis). Bueno submits that 'representation is an *intentional* act relating two objects' (2010, 94, original emphasis), and Bueno and French point out that using one thing to represent another thing is not only a function of (partial) isomorphism but also depends on 'pragmatic' factors 'having to do with the use to which we put the relevant models' (2011, 885).

According to this way of thinking then, it is the activities of model users, in offering theoretical hypotheses proposing resemblances, that answer the Semantic Question. So how does PROPOSED RESEMBLANCE fare with respect to the conditions of adequacy offered in the introduction? Pretty well, as it turns out. The Surrogative Reasoning Condition is met via that fact that in proposing a resemblance, a model user can infer from the fact that a model has a certain feature (structural or otherwise) to the claim that the target shares that feature. And since theoretical hypotheses can be true or false, it allows for the fact that the target might fail to have that feature, thereby accommodating cases of misrepresentation.

PROPOSED RESEMBLANCE also avoids at least some of the issues with the logical properties of resemblance discussed earlier: people typically don't offer theoretical hypotheses according to which some object resembles itself, so reflexivity is avoided. Whether it avoids the issue arising from symmetry is less clear. Plausibly, proposing that x resembles y seems to engender a commitment that y resembles x, but for our current purposes we'll put this aside.

Moreover, the accidental resemblances that featured in Putnam's thought experiment no longer pose a problem for this way of thinking about representation: the ant's trace doesn't represent Churchill because no one, let alone the ant, proposes that the two are similar. All in all, this way of thinking about the Semantic Question seems like it has potential, although it is worth nothing that, as with RESEMBLANCE, PROPOSED RESEMBLANCE has nothing to say about target-less models.

One of the main arguments in its favour is how PROPOSED RESEMBLANCE accommodates surrogative reasoning: a model user can exploit a proposed resemblance to infer that a target has a feature from the fact that the model does. But a question remains whether this suffices to accommodate all cases of surrogative reasoning. Giere suggests that he's open to other ways when he describes how models allow for such reasoning: '[o]ne way, perhaps the most important way, *but probably not the only way*, is by exploiting similarities

between a model and that aspect of the world it is being used to represent'
(2004, 747, emphasis added). Giere does not expand on what other ways he
has in mind, but given that structural relations like isomorphism play no role
in his discussion, it is unlikely that this is what he had in mind. If so, this
amounts to the admission that there are kinds of representation that are not
based on resemblance. We agree, and we will encounter cases of this kind in
the next section and in Section 4. But if there are such forms of representation, then PROPOSED RESEMBLANCE cannot be a complete answer to the
Semantic Question.

2.4 The Accuracy Question

Can resemblance be invoked to answer the Accuracy Question? The way in
which resemblance (of relevant features, structural or otherwise) could be so
used should now be clear: in proposing that a model resembles a target with
respect to some feature, a model user establishes that the model represents the
target as having that feature; if the target does in fact resemble the model with
respect to that feature, then the model accurately represents the target with
respect to that feature.

Such an answer meets our success conditions introduced in the introduction.
The fact that context is required in specifying the features relevant to the
resemblance means this answer meets the Contextuality Condition.[18] In
a context where the Ptolemaic model is proposed to be similar to the solar
system only with respect to how the celestial bodies appear in the sky, the two
are so similar, and thus the model is accurate with respect to that feature. But in
a context where the model is proposed to be similar to the target more generally,
including with respect to features like the trajectory of the orbits themselves, the
two are not so similar (since in the model the planets orbit the earth, and in the
actual solar system they all orbit the sun) and so the model is inaccurate in those
respects.

This answer also meets the Gradation Condition, in fact in multiple ways.
A model user may propose more and more relevant similarities between a model
and a target, allowing for more and more accurate representation if those
resemblances hold.[19] Alternatively, if a model user proposes some fixed collection of relevant features, the extent to which the model actually resembles its
target can also come in degrees, either in the sense of resembling the model only
with respect to some of them or in terms of the importance of the resemblances

---

[18] If resemblance is explicated in terms of structure, model users may propose different mappings
(*cf.* footnote 8).
[19] If resemblance is explicated in terms of structure, they may include a larger number of relations
in the structure or propose more demanding mappings (again *cf.* footnote 8).

in question or in terms about how similar the model and the target are with respect to those features, when similarity is understood at the level of the features themselves (*cf.* the discussion of imprecise features in Section 2.1). So far so good.

The question then is whether proposed resemblance underpins all instances of successful surrogative reasoning – this is the issue that arose at the end of the previous section. If it's not, then resemblance cannot be a complete answer to the Accuracy Question. As we've already indicated, we think that there is reason to doubt that resemblance can play this kind of universal role. Let's start with an easy (and by now well-rehearsed) example before discussing a bona fide scientific case.

Consider the London Tube map. It's a two-dimensional array of dots, representing tube stations, connected by different coloured lines, representing tube lines. Anyone with some familiarity with such maps finds it easy to use the map to navigate the underground system. A user identifies the dot representing the station where they'll begin their journey, the dot representing their desired destination station, and then they can determine various different possibilities for which lines they should take, and where they need to change. For example, if you want to travel from Leyton in East London to Victoria in Central London, then you can take the route represented by the red (Central) line connecting the dot representing Leyton to the dot representing Oxford Circus, followed by the light blue (Victoria) line connecting the latter to the dot representing Victoria. Using the map to plan such a journey is exactly what we mean by 'surrogative reasoning'.

What relationships between the map and the underground system does the map user exploit when performing such reasoning? It is commonplace to point out that the tube map represents the *topology* of the underground (i.e. the way in which the stations are connected to each other).[20] Moreover, the topology of the map is the same as the topology of the tube system, and in this sense, when model users reason about topological features of the latter by investigating the former, they exploit this resemblance, something that PROPOSED RESEMBLANCE captures nicely. But there are other relationships that are important too. A map user also cares about which specific coloured lines represent which specific underground lines, which is not a topological feature. They know that the red line represents the Central line, that the light blue line represents the Victoria line, and so on. So when the traveller was planning their trip from Leyton to Victoria, they knew more than 'I need to change at Oxford Circus', they knew

---

[20] It is also commonplace to point out that it doesn't represent the *metric* (i.e. the distance between the stations).

that they needed to change *onto the Victoria Line* at Oxford Circus. To that end they need to understand the colour coding of the tube lines – specifically, they need to know that the light blue line represents the Victoria Line. But the relationship between colours on the map and underground lines in the world are purely conventional, and there's no sense in which the colour light blue is co-instantiated by, or similar to, the features that distinguish the Victoria Line from others. So this style of surrogative reasoning doesn't seem to proceed via exploiting resemblances.[21]

Conventional elements in representations are not limited to mundane representations like tube maps; they are also used in proper scientific contexts. As an example, consider fractal geometry. Most readers will be familiar with colourful pictures of the Mandelbrot set, which have become so popular that one can even find them printed on T-shirts.[22] The function of the colours in these pictures goes beyond the aesthetic. In fact, they are a colour code. One starts by considering an iterative function that takes as input parameter a complex number c. One then asks, for a particular number c, whether the function converges or diverges, and if it diverges, how fast it does so. The result of this is then colour coded: if the function converges for c, then the point in the plane representing c is coloured black, and if the function diverges, then a shading from yellow to blue is used to indicate the speed of divergence, where yellow is slow and blue is fast (Argyris et al. 1994, 663). Of course, a different colour coding can be used. The association of certain colours with certain speeds of divergence is entirely conventional. Nevertheless, the picture provides important scientific information, and it does so without invoking any similarity between the representation and the target – divergence speeds are not similar to colours!

What this suggests is that there are some forms of surrogative reasoning that exploit non-resemblance relationships between representations and their targets. In this case, a particular feature in a map or the fractal image (i.e. a colour) is readily understood as representing some distinct feature in its target (i.e.

---

[21] One could respond here that what's important is just the *distinctness* of the lines: as long as different colours represent different lines, there's a more abstract feature shared by the map and the world that is being exploited (e.g. there being three different colours passing through the dot marked 'Oxford Circus', and there being three different lines passing through the station itself). However, this fails to accommodate the fact that the map user knows she has to change *onto the Victoria Line*, not that she just has to change onto *some* line (and it's important that she knows this, otherwise she might end up on the Bakerloo line). The relationship between colours and tube lines is more specific than just different colours represent different lines; it explicitly pairs up a colour with a line, and a map user exploits this pairing up in their reasoning with the map. For further discussion of maps in this context, see Nguyen and Frigg (2022).

[22] One of these pictures can be seen here https://commons.wikimedia.org/wiki/File: Mandelbrot_set_10000px.png

a tube line or a speed of divergence), but without invoking a resemblance relation. Here the model–target feature associations that are relied upon are based on convention. In other cases the underlying association may be more complex. The ship example in the introduction illustrates this. We will go over the details in Section 4, but it pays to have a brief look at some details of the case now to see that in order to perform (successful) surrogative reasoning, a model user should associate model features with features of their targets in a way that doesn't rely on resemblance alone.

Recall the scenario. You're a shipbuilder tasked with determining what size engine a ship needs to be able to travel at a certain speed. You have built a scale model of the ship, which is the same geometric shape as the ship, but is much smaller. Suppose that the model is a $1:s$ scale (e.g. $1:100$) of the full-sized ship, which means that each of its linear dimensions (e.g. length, width, and height) are all $1/s$ the size of the full-sized ship. You tow your scale model through a tank and measure the complete resistance, R_C, that it faces, which in turn allows you to calculate how big an 'engine' (i.e. how much force you need to tow it with) the scale model ship needs to move at various speeds. The question then is how can you translate the results of your investigation into something that's informative about the full-sized ship itself?

The first thing you might ask is: if you performed your experiment on the model ship at various velocities, which one is relevant for determining the behaviour of its target? A defender of resemblance might think: the same velocity at which the full-sized ship will travel. This turns out to be wrong. What's important when performing these model experiments is not that the model and the target move at the same velocity, it's that they create the same wave pattern, and when the two are different sizes, this requires different velocities. We'll go into more detail about this in Section 4, but for now what's important is that in order to ensure that the model ship creates the same wave pattern as the full-sized ship, the relevant velocity needs to be scaled by the square root of the ratio of their lengths.

The next question is: if we determine the wave-making resistance faced by the model ship, how do we translate that to the wave-making resistance faced by the full-sized ship? A naïve defender of resemblance might say: we need to scale the resistance proportionally to the scale s of the model. This is wrong again. As it turns out, the way in which wave-making resistance changes across scales depends on the ratio of the *cubes* of the linear dimensions, that is, the ratio of the amount of water displaced by the model ship and full-sized ship (again, we'll go into more detail about this in Section 4).

So in order to perform the right model experiment, and in order to translate its results into true claims about the target ship, the shipbuilder needs to ensure that the model and the target do *not* resemble one another with respect to their velocities (not even in the imprecise way discussed earlier in Section 2.2) and needs to scale the wave-making resistance faced by the ship in a way that depends on the ratio of the cubes of their lengths. The point then is that in order for the model ship to accurately represent the full-sized ship, the shipbuilder needs to perform some relatively complex calculations on the results of the model experiments, calculations that rely on different scales for different values (e.g. velocity and wave-making resistance) and that require that the model and the target drastically diverge from one another with respect to at least some features (e.g. velocity). This kind of reasoning does not seem to proceed via merely proposing that the model and the target resemble one another in some relevant respects, with the model being accurate if they are so similar.

At this point a defender of resemblance might object. In setting up the reasoning this way we're ignoring the fact that it must be the case that the model resembles the ship with respect to its shape, and the fact that the reason why the velocities need to diverge from one another is to ensure that the two resemble one another with respect to their wave-making patterns. Indeed, part of the reason why scale models more generally fail to resemble their targets in certain respects is to ensure that they do thereby resemble them in some other particularly important respect.[23] And they are right: there is a sense in which these resemblances do play *some* role in the shipbuilder's successful surrogative reasoning. But our point is that the way the shipbuilder reasons with the model is much more complex than simply proposing resemblances: you need to scale your measurements, distort some features, and work with complex scaling factors that directly connect features of the model with features of the full-sized ship that do not, in any meaningful sense, 'resemble' each other. Of course, after the fact, one may be able to identify some relevant set of features that the model and the target share (including the shape of the waves they produce), but this tells us nothing about the way in which the shipbuilder's reasoning proceeds, and nor should we think that these shared features are the only ones that they care about. At this point, labelling the reasoning process as one where 'resemblances' are exploited is uninformative: it's a label attached after the fact. What matters are the specific ways in which the shipbuilder associates features of the model with explicitly different features of the target, and, as we will see in Section 4, we can develop answers to the Semantic and Accuracy questions that build these associations directly into an account of representation in a way which is both more liberal than PROPOSED RESEMBLANCE, and more informative of the way in which the shipbuilder actually reasons.

---

[23] It is often required that scale models resemble their targets with respect to certain 'dimensionless parameters', like their Froude number. For a philosophical discussion, see Sterrett (2009).

2.5 The Model Question

Even if the aforementioned worries can be addressed, there remain further problems facing PROPOSED RESEMBLANCE. These arise from the restrictions that such an account puts on the nature of models, and targets, themselves. We begin this section with a more general discussion about how to conceptualise models on this account, and then turn to the restrictions.

So far we have remained relatively silent about what models are, ontologically speaking. In the case of the ship model this seems straightforward: the model is the concrete system consisting of the ship in the tank. But what about the car jump model, or more generally, non-concrete models, which might represent concrete systems in the world, but are not themselves concrete?

In Section 1.3, we introduced the notion of a model-description. It's now useful to distinguish between the *model-description* and the *model-system*. In the case of the car jump model, the former contains all the information used to specify the car jump model and the associated mathematical equations that go along with it (introduced in Section 1.1). The latter is the system specified by this description. What's crucial to note is that this system is not the actual car-and-bridge system itself; it's some other system (the model-system) that in turn represents the car-and-bridge system. This is required if we are to understand the idea that the model resembles its target in the first place: it doesn't make sense to ask whether or not a model-description resembles its target since descriptions are in plain English and equations don't have the sort of features that are relevant in the modelling context (equations themselves don't have trajectories, or fall through the air). So it's the model-system that may be said to resemble the actual car, not its description.

This is in line with Weisberg's (2007) distinction between direct and indirect representation, which is now generally accepted. Weisberg submits that modelling is different from other kinds of scientific reasoning precisely because of its indirect nature.[24] When a scientist builds a model, they construct a model-system, which then becomes the focus of study. They investigate how it behaves and what features it has, and then assesses the system's relation to the intended target. At this stage, according to Weisberg, similarity becomes crucial because if the model-system is deemed 'sufficiently similar to the world, then the analysis of the model is also, indirectly, an analysis of the properties of the real-world phenomenon' (*ibid.*, 209–10). The same applies to those who prefer to explicate resemblance in terms of isomorphisms. A system of equations isn't the

---

[24] For further discussion of modelling as indirect representation, see, for instance, Godfrey-Smith (2006) and Knuuttila (2021). The idea is also discussed in Section 4. For opposing viewpoints, see Toon (2012) and Levy (2015).

sort of thing that can enter into an isomorphism with a target system in the world. Equations specify a mathematical structure, and that structure is the object that can then be related with the system in the world through an isomorphism. In the case of the equation in the car model, the equation specifies a particular geometrical trajectory in the x and y coordinates and it's this structure, not the equation, that represents the car jump.[25]

This way of conceptualising models also fits with scientific practice, where scientists often talk about the features and dynamics of their models as if they were objects of sorts. The question that arises here then is how we should think about the ontological status of model-systems. This is a significant question, and confronting it in detail would take us deeper into the territory of metaphysics than we can venture here. But what we can ask is: What restrictions do answers to the Semantic Question and the Accuracy Question put on accounts of the ontological status of models and their targets? The problem is that the requirements of PROPOSED RESEMBLANCE (and variants thereof) are difficult to accommodate. On the one hand, similarity accounts require that models are similar to their targets with respect to material features, and therefore need to be the sorts of things that can have such features. On the other, structuralist accounts require that target systems can be isomorphic to their models, and therefore need target systems to be the sorts of thing that can be related to structures via structure preserving functions. Neither of these restrictions is straightforward.

Take similarity first. In the case of the ship model, presumably the relevant features include material properties and relations like the resistance faced by the ship and its model. This is fine, the ship model is a concrete object, and it does literally have such features. But things are more problematic for non-concrete models. When using a Newtonian model to coordinate a car stunt, the relevant features would presumably include things like the masses of the sphere in the model and the real Aston Martin in the world, the trajectories of both objects, and the forces acting on them. The problem then is that these features are typically *material* features. But non-concrete model-systems, if they exist at all, don't seem to be the sorts of things that can have such features. The car jump model doesn't exist in the concrete actual world, and when we investigate it we don't prod, push, or pull it; we investigate it through thought, not physical interaction. The model isn't the sort of thing that we can probe with physical instruments. It's an abstract object. But then it's difficult to understand what makes it true that the sphere in the model is affected by gravity, since gravity doesn't affect abstract objects.

---

[25] Indeed, it is one of the core posits of the so-called semantic view of scientific theories that structures themselves, and not the varying descriptions that we can give of these structures, are the units that represent a theory's target systems. See, for instance, van Fraassen (1980).

The same goes for all kinds of models where (i) according to the modelling context it's material features that we're interested in, and thus it's material features that models must have if they are to be similar to their targets with respect to those features, and (ii) the model in question is a non-concrete model and therefore isn't the sort of thing that can have such material features. Appealing to similarity as underpinning representation (accurate or otherwise) seems to require that non-concrete models be the sorts of objects that can have material features, and yet these model-systems can't be identified with anything concrete in the world. So the question facing such a view is: how can non-concrete model-systems be similar with respect to material features in the first place?[26] In this way, PROPOSED RESEMBLANCE has implications for the Model Question that pose a challenge to the Model-Truth Condition.

When it comes to accounts of representation that rely on structural relations between models and their targets, the restrictions are reversed. When resemblance was explicated via similarity, the restrictions required that the *model* have material features; when it comes to structural accounts, the restrictions require that (both the model and) the *target* be a structure.[27] In order for a system to enter into an isomorphism, it has to be a structure in the sense outlined in Section 2.2. And this requires that it be a set of elements with properties and relations extensionally defined over them. But scientific models represent physical systems, not mathematical structures. The car jump model represents a car jumping over a bridge, not a structured set. Prima facie at least, it is a category mistake to assume that a target system is the sort of thing that can enter into an isomorphism in the first place. Thus, the onus is on defenders of the idea that representation can be analysed in terms of isomorphism (or other mappings) to provide an account of how target systems can enter into such mappings.

Such an account could be provided in various ways. In the first instance, one could argue that scientific models represent *data models* rather than target systems. Data models result from first performing various measurements on a target; then processing and cleansing the results of these measurements; and finally presenting them in a mathematically regimented form. A simple example of a data model is a curve that is fitted through the experimental data points. Data models can be thought of as structures. Let us apply this idea to the car

---

[26] For more on this problem, see Thomson-Jones (2010), and for possible responses, see Teller (2001) and Thomasson (2020).

[27] In contexts where the model is a non-concrete mathematical model, this might be straightforward. But in what sense should we think about the concrete ship being towed through a tank as a mathematical structure? Presumably, whatever sense we can make of the idea that target systems are structural must also be put to work in making sense of the idea that concrete models are structural.

jump model: suppose that in your role as the stunt coordinator, you were to adopt a coordinate system applied to the bridge system, and then, as the performers drove over the bridge completing the stunt, you measured the location of the car with respect to the x and y, coordinates of the centre of the car at consecutive instants of time. The result would be a sequence (indexed by time) of pairs of real numbers (x, y). The data thus gathered are called the *raw data*. The raw data then undergo a process of cleansing, rectification, and regimentation: you throw away data points that are obviously faulty, you take into consideration what the measurement errors are, and you finally replace discrete data points by a continuous function. In this instance, your continuous function would be defined on the real numbers, and would be the sort of thing that could be isomorphic to the model you were using to represent the trajectory in the first place.

But here one should object: the model doesn't represent the data model extracted from measuring the car's location, it represents the car itself.[28] In general, models can represent systems that we have and will never measure; and for systems from which we have extracted data, we shouldn't conflate the extracted data with the target system itself. Bogen and Woodward (1988) illustrate this difference as follows. Suppose you're investigating the melting point of lead. You might carefully heat various samples of lead and measure the temperature at which they undergo a phase transition from solid to liquid. Depending on the details of the experimental set ups (the thermometer used, the instant at which the temperature is recorded, and so on) these measurements may disagree with one another to some extent. You may then take the average of your measurements as the melting point, the target phenomena you are inter-ested in, even if this average value doesn't correspond to any particular meas-urement. This average value is your data model. The point then is that any theoretical model of the phenomena is ultimately targeted at the melting point, not any of the measurements thereof (either individual or processed). So arguing that scientific models represent data models seems to give up on the idea that they represent physical phenomena in the world.[29]

An alternative way of conceptualising 'the structure' of target systems is to locate the structure 'in' the target systems themselves. One way of doing this would be to adopt a Pythagorean position about the ontology of the universe: the

---

[28] Note that this worry remains even if one grants that the data model represents the target system: that a model represents a data model, which in turn represents a target system doesn't entail that the model represents the target without the additional assumption that scientific representation is transitive, which is unmotivated. See Brading and Landry (2006) for further discussion.

[29] Van Fraassen (2008) provides an account of target-end structures that accepts the distinction between data and target, but collapses its relevance in the context of scientific representation. See Nguyen (2016) for a critical discussion.

world simply *is* a mathematical structure (Tegmark 2008). This is a radical suggestion, and we assume that most would be disinclined to take it too seriously. A more reasonable option would be to accept that target systems are physical systems, but then to argue that they have parts that can be identified with elements of a set, giving a domain of a structure. The physical properties and relations of those parts have extensional counterparts then delivering the requisite extensionally defined properties and relations required of a structure. As a result, the physical system can be thought of as *instantiating* the resulting structure. The idea that physical systems can instantiate structures in this way is then entirely analogous to the way we think of physical objects as instantiating properties (or 'universals'). Just as the London bus instantiates a particular shade of red, the family under the *is a descendent of* relation presented in Section 2.2 instantiates a structure consisting of three objects and a two-place relation. As Shapiro puts it: 'the problem of the relationship between mathematics and reality is a special case of the problem of the instantiation of universals. Mathematics is to reality as universal is to instantiated particular. As above, the "universal" here refers to a pattern or structure; the "particular"' refers not to an individual object, but to a system of related objects. More specifically, then, mathematics is to reality as pattern is to patterned' (1983, 538).

This way of identifying a target-end structure faces objections analogous to the 'spurious similarities' objection discussed in the context of sharing irrelevant material features. Depending on which physical features are taken as relevant in a given context, target systems instantiate different structures, and as a result, a given model may be isomorphic to, and thereby accurately represent, the target in one context but not in another. By way of illustration, consider again our family of three. In one context one can focus on the relation *is a descendent of*, leading to structure with a binary relation and no one-place property. In another context, one can consider the property *plays a musical instrument*, resulting in a structure involving three objects with a one-place extensional property. These structures are clearly not the same, and yet both are structures of the family. So there is no such thing as *the* (one and only) structure of a target that the model could relate to without further explication. This problem is general because any system can be divided into 'parts' in different ways, delivering different domains, which provides yet another way in which 'the' target-end structure is underspecified. In fact, the objection is even more damning, since the parts of the structure can be collected into arbitrary collections, providing arbitrary extensional properties and relations for a structure.[30]

---

[30] For further discussions of different ways of dividing a system in to parts, see Nguyen and Frigg (2021) and Pincock (2012, Ch. 2). See also Frigg and Votsis (2011), who discuss this problem in the context of the so-called 'Newman's objection' to structural realism.

So the idea that targets instantiate structures faces an underdetermination objection: depending on how one specifies the parts of the target and the physical features of the system, one arrives at any number of structures instantiated by the target. Which of these are supposed to be related to the structure being used to represent the target? A plausible suggestion here is that it's the one that is *ascribed* to targets by model users in contexts of representation. The idea here is that a model user specifies, in a physical vocabulary, how the target is to be decomposed into parts and which physical features are relevant. This specification then provides a structure which one can compare to the mathematical structure being used to model the target. This seems like a viable strategy, but notice that the activities of model users, and the conceptual schemes they employ in describing a target, are accorded a central role in the process. Accordingly, producing a representation involves much more than just an isomorphism; it involves conceptualising a target in certain way and ascribing a structure to it.[31] There is no such thing as an account of representation that only involves structural terms, and a definitive formulation of such an account of representation would have to make the involvement of these further elements explicit.

This completes our discussion of using resemblance to account for scientific representation. On either way of explicating it, in terms of material or structural features, we have argued that the most viable way of doing so requires invoking *proposed* resemblances to answer the Semantic Question, and using actual resemblances to answer the Accuracy Question. But such an approach still faces the challenge of accounting for cases of surrogative reasoning where model-users exploit model–target relationships without invoking resemblances, those that, for example, involve conventional ways of encoding information, or that rely on how background theories apply to systems that don't resemble one another in any obvious way. Moreover, the constraints that resemblance accounts put on answers to the Model Question provide further worries: they are incomplete without an explanation of either how non-concrete models can be similar to their targets with respect to material features (features that non-concrete objects don't seem to have, at least prima facie), or how physical target systems can 'exhibit' the structures required in order for them to be isomorphic to their models.

3 The Inferential Conception

3.1 Introduction

So far we have treated the Surrogative Reasoning Condition as something to be explained by an answer to the Semantic Question. In the accounts discussed in

---

[31] This is acknowledged by modern defenders of versions of PROPOSED RESEMBLANCE, e.g. van Fraassen (2008), Bueno (2010), and Bueno and French (2011).

the previous section, the idea was that invoking model–target resemblance relations provided an explanatory basis for surrogative reasoning: a scientist can use a model to reason about its target *because* they are working under the assumption that the two resemble one another. However, as we have seen, whilst these relations may play such a role in some cases, they do not provide a universal explanation of the ways in which scientists reason about their targets via investigating their models. In light of this, one way to proceed would be to try and find some other kind of model–target relation that would provide such explanations. But there is another option available: rather than trying to explain surrogative reasoning via some other kind of model–target relation, we could instead take such reasoning as philosophically *primitive*. This approach has become known as 'inferentialism' (because in performing surrogative reasoning model users draw inferences from the model about the target). We explore this approach in this section.

We begin with a discussion of an answer to the Semantic Question motivated by this approach, and sketch how it fares with respect to the other questions and conditions that structure this Element (Section 3.2). We then turn to the approach itself and attempt to motivate it via discussions drawn from elsewhere in philosophy: inferentialism in the philosophy of language and deflationary theories of truth. We argue that these discussions fail to appropriately motivate such an answer to the Semantic Question (Section 3.3). We then turn to a discussion of alternative answers which are, at least in part, reactions to the inferential approach developed thus far (Section 3.4). We conclude that there remain significant open questions for inferentialist approaches.

3.2 Representational Deflationism

Taking surrogative reasoning as philosophically primitive amounts to giving up on the idea of explaining it in other terms. One can simply assume that scientists use models to draw inferences about their targets, and deny that this is to be further explained, at least at a general level, by a philosophical theory of scientific representation. This is at the heart of Suárez's 'inferential conception' of scientific representation (2004, 2015), the first explicitly inferentialist position in the recent debate over scientific representation:

> INFERENCE: a model represents its target if and only if (i) the representational force of the model points to its target; and (ii) the model allows competent and informed agents to draw specific inferences about the target.[32]

---

[32] Suárez (2004, 2015) offers conditions (i) and (ii) as necessary but insufficient conditions on scientific representation and claims that we shouldn't expect to find sufficient conditions. However, he does consider the idea that they could be taken to be sufficient but

Condition (ii) in effect says that models allow for surrogative reasoning, and this is thus assumed (rather than accounted for) by INFERENCE (we will discuss what it means for something's 'representational force' to 'point' at something else below). As such, you might immediately object that INFERENCE cannot be an acceptable answer to the Semantic Question because the conditions don't provide any purchase on *how* models inform us about their targets. But this is by design: the conditions are not supposed to be explanatorily. As Suárez puts it: we should seek 'no deeper features to representation other than its surface features' (2004, 771), or 'platitudes' (Suárez and Solé 2006, 40). It's this aspect that involves taking surrogative inference as philosophically primitive. Before discussing this strategy in detail, it's worth briefly discussing the conditions themselves, and clarifying how they help answer the questions and conditions we introduced in Section 1.

Take 'representational force' in condition (i) first. Intuitively at least, this condition allows INFERENCE to meet the Directionality Condition. That the representational force of x points towards y does not entail that y performs a representational role at all, let alone that its representational force points back towards x. What exactly 'representational force' amounts to in detail is, however, not further explicated, since, again by design, it is supposed to be a 'surface feature'. However, Suárez does provide two comments that help us get more of a handle on the notion. First, it is 'satisfied by mere stipulation of a target for any source' (2004, 771) (where he uses 'source' as a general term to refer to epistemic, including scientific, representations).[33] Second, whilst denotation (either in addition to, or established by, stipulation) might be sufficient to establish representational force, it should not be identified with it. This latter point is motivated by two observations. For one, as we discuss in more detail in the next section, for x to denote y, y has to exist. Thus, if one were to include denotation as a necessary condition for scientific representation, then one would rule out models that don't represent any actual target. As such, one would violate the Targetless Models condition, something that Suárez is disinclined to do (2015, 44). So 'representational force' allows for models that don't represent any actual target in a way that denotation does not. For another, denotation is a substantial relation; lots can be (and has been) said about what it takes to establish it. As such it is not a 'surface feature', and therefore goes against the deflationary motivation for INFERENCE. So the condition ensures that

---

non-explanatory (see in particular Suárez and Solé (2006, 41), and Suárez (2015, 46)), and it is their lack of explanatory force that is our focus here. As such we set the question of sufficiency aside in this section, but see Frigg and Nguyen (2020, Ch. 5) for a discussion.

[33] Callender and Cohen (2006) also invoke stipulation in the context of scientific representation. For extended discussions of their approach, see Boesch (2017), Frigg and Nguyen (2020, Ch. 2), and Ruyant (2021).

INFERENCE meets both the Directionality Condition and the Targetless Models Condition, but beyond that, in line with the motivation of the account, little else can be said.

Similar observations apply to INFERENCE's condition (ii), that a model must allow agents to draw inferences about the target. As noted earlier, this is synonymous with surrogative reasoning, and as such INFERENCE meets that condition by design. Moreover, Suárez is explicit that it 'accounts for inaccuracy since it demands that we correctly draw inferences from the source about the target, but it does not demand that the conclusions of these inferences be all true, nor that all truths about the target may be inferred' (2004, 776). To illustrate: it is correct to infer from the Ptolemaic model of the solar system that the various celestial bodies orbit the earth in complex epicycles, but this model-to-target inference does not yield a true conclusion. It is also correct to infer from a heliocentric model that the earth orbits the sun, and in this case the conclusion is true, but the model may not yield all truths about the solar system (e.g. it does not represent the orbital inclination of the planets' orbits). Thus, INFERENCE meets the Misrepresentation Condition.

Here, we also see how INFERENCE addresses the Accuracy Question. When the conclusions of the model-to-target inferences are true, the model is an accurate representation with respect to the features of the target it describes. If you infer from the ship model that SS *Monterey* will face such and such resistance when sailing at a given speed, and this is true, then the ship model is an accurate representation with respect to the resistance faced by the actual ship at that speed. If you infer something false, then the model is inaccurate in the relevant respect.

This way of thinking meets the Gradation Condition: the more of the conclusions are true, the more accurate the representation is (and this can be further nuanced in terms of weighing the conclusions according to their relative importance if desired). In addition, whilst condition (ii) of INFERENCE requires the drawing of specific inferences from model to target, it doesn't rule out the conclusions of these inferences being imprecise (e.g. perhaps all you infer is that the ship will face a resistance that lies in the interval $[R - \epsilon, R + \epsilon]$, and you can't be more specific than this). So one can also make sense of the gradual nature of accurate representation in terms of true conclusions that are more or less precise.

Given that INFERENCE offers no detail about the model-to-target inferences users can draw, it also has the freedom required to meet the Contextuality Condition: if the purpose for which a scientist is using a model requires that the model be accurate in some respects rather than others, then the scientist can require that conclusions drawn about the former, but not the latter, be accurate.

The addition of the clause 'competent and informed agents' offers another way of thinking about the contextual nature of model accuracy. Different contexts may set different requirements on what it takes for an agent to be competent and well informed, depending on the nature of the model, the purposes to which it is put, and the inferences to be drawn.

Finally then, INFERENCE chooses to say little about the Model Question. All that is required is that models have an internal structure that allows agents to draw inferences about the target, and given that Suárez illustrates the account with examples as diverse as a graphical representation of a bridge (2015, 42), a sheet of paper and two pens (2004, 772), a mathematical equation (2004, 774), mathematical structures in the sense discussed in the previous section (Suárez and Solé 2006, 43), and paintings (2004, 777), it is clear that INFERENCE places few, if any, constraints on answers to the Model Question.[34]

So, INFERENCE, by design, meets the conditions required of answers to the Semantic Question and the Accuracy Question, and remains silent about the Model Question. However, the way in which it does so is, in a sense, trivial, since the conditions it offers are deliberately non-explanatory surface features. In effect, INFERENCE takes the Directionality and Surrogative Reasoning Conditions, and simply defines a scientific representation as something that behaves as the conditions require. This may seem unsatisfactory. As Contessa puts it:

> On the inferential conception, the user's ability to perform inferences from a vehicle [e.g. a model] to a target seems to be a brute fact, which has no deeper explanation. This makes the connection between epistemic [e.g. scientific] representation and valid surrogative reasoning needlessly obscure and the performance of valid surrogative inferences an activity as mysterious and unfathomable as soothsaying or divination. (2007, 61)

This concern is also shared by Chakravartty, who argues that views like INFERENCE need to be combined with accounts that emphasize 'objective' model–target relationships, for:

> in the absence of substantive relations of similarity (in some form or other) between scientific representations and their targets, it would be something of a mystery how these devices represent things in scientifically interesting ways at all. (2010, 203)

---

[34] For an inferentialist account of the application of mathematical models, see McCullough-Benner (2020).

But what motivates INFERENCE is precisely the idea that representational force and surrogative reasoning are brute facts: they are deliberately taken as philosophically primitive.

3.3 Why Be an Inferentialist?

Is it legitimate to take surrogative reasoning as conceptually primitive? When performing philosophical investigations into the ways in which scientists draw inferences about their target systems by investigating the behaviour of their models, we are trying to find out why and how they draw such inferences. Surrogative reasoning is something to be explained, and simply taking such reasoning as primitive gives up on the idea that such an explanation can be given. Proceeding in this way thus seems to amount to abandoning the philosophical project of understanding how scientific representation works.

Whilst we think there is something to the objection, it's worth probing it further by looking at how inferentialism can be justified. There are two ways of bolstering support for an inferentialist approach to scientific representation: looking at parallel movements in the philosophy of language and in theories of truth.

Let us start with the philosophy of language, where there is a rich intellectual tradition that self-identifies as 'inferentialist'. Classic statements include Brandom's (1994, 2000); for a useful introduction see Murzi and Steinberger's (2017), whose discussion we follow here. Surprisingly, the parallel with inferentialism in the philosophy of language has garnered little attention among philosophers of science,[35] and so it is worth exploring whether drawing this parallel can dispel the concern that INFERENCE leaves scientific representation 'mysterious'.

In the context of an analysis of language, we typically distinguish between the meaning of sentences and other linguistic objects ('semantics'), the conditions under which they are used, or uttered, by speakers of that language ('use-conditions'), and the explanations of these features of language ('meta-semantics').

Semantic analyses in linguistics and the philosophy of language provide theories with which we can assign meanings to sentences. Standard referential semantics proceeds by analysing referential relations between parts of the sentence and features of the world, and then understands the meaning of sentences as being somehow composed of these referential relations. For example, consider the sentence 'the dog is on the bed'. Its meaning – that the

---

[35] However, see de Donato Rodriguez and Zamora Bonilla (2009), Kuorikoski and Lehtinen (2009), and Khalifa, Millson, and Risjord (2022) for efforts in this direction.

dog is on the bed – is assigned to it due to the expression 'the dog' referring to the dog, the expression 'the bed' referring to the bed, and the expression 'is on' referring to a relation that holds between two objects if the first is on the second. Taking these three referential relations together, combined with the order of the phrases in the sentence in which they occur, delivers the following: 'the dog is on the bed' means that the object referred to by 'the dog' stands in the relation referred to by 'is on' to the object referred to by 'the bed'.

We can also analyse the conditions under which the sentence 'the dog is on the bed' is uttered by speakers of English. One such analysis would start from the idea that the sentence is, or should, be used when the dog is, as a matter of fact, on the bed. But we could also analyse the relationship between uses of that sentence, in relation to uses of other sentences. For example, 'the dog is on the bed' is an appropriate response to a question like 'where is the dog?'. More generally, we can focus on the inferential relationships between sentences: given 'the dog is on the bed' one can assert that 'there is an animal in the house', 'the dog isn't in the park', and so on.

Meta-semantic analyses give us theories about why sentences mean what they do, and why sentences are used in the contexts in which they are. Non-inferentialists explain why sentences are used when they are by appealing to the underlying referential semantics. The latter explains the former. In particular, these semantics can be invoked to explain inferential relationships (i.e. one can infer from 'the dog is in the house' that 'there is an animal in the house' because of the relationship between the semantics of 'dog' and 'animal'). Inferentialist approaches to meta-semantics, however, reverse the order of explanation: referential semantics doesn't explain the inferential roles of sentences; the inferential roles of sentences explain the referential semantics.[36] According to the inferentialist, their opponents take for granted referential, or more generally semantic, relationships and use these to explain inferential ones, and they suggest that, in contrast, one can take inferential relationships for granted, and use these to explain the semantic ones. In these terms, the inferentialist project is primarily one in meta-semantics: take the inferential role of sentences as basic, and then use them to *explain* the semantics. Indeed, Murzi and Steinberger (2017) emphasise the importance of understanding inferentialism in this way, particularly the fact that as a meta-semantic doctrine it is compatible with standard referential semantics.

What this detour into the philosophy of language demonstrates then is that taking inferential relationships as conceptually primitive is a live option. The

---

[36] The inferentialist could also adopt a non-standard account of the semantics. See Murzi and Steinberger (2017) for a discussion. A particularly relevant non-standard account would be deflationary semantics (Brandom 1994, Ch. 5), which we discuss later in the text.

inferentialist in the philosophy of language doesn't do anything they're not entitled to; their approach is to assume inferential relationships at the meta-semantic level in order to explain the semantics, and this is a viable option. This leads to a philosophical debate about the explanatory priority between inference and semantics. On one side, non-inferentialists help themselves to some basic semantic notions (e.g. reference) and then put these to explanatory work in order to help us understand the use, and inferential role, of sentences. In contrast, inferentialists help themselves to some basic notions concerning use (e.g. the inferential role of sentences) and then put these to work in order to help us understand semantic notions like reference. Both take *something* as primitive and use it to explain something else; they just disagree about the order in which this should be done.

How does this carry over to the context of providing a philosophical analysis of scientific representation? Prima facie at least, it seems to help legitimize the idea that we can take the inferential role of scientific models as philosophically primitive. But can we do more than gesture at the analogy to defend the approach discussed in this section?

The first thing to note is that this reversal of the order of explanation isn't present in discussions of INFERENCE. If we understand inferentialism in the philosophy of language in the way described above, the analogy would motivate the idea that we can take model-to-target inferences as primitive and then use this to develop an account of the semantics of models, that is, the model–target relations that are explained by the inferential role that models play.[37] Inferentialism could be taken as the framework in which prototypical models are analysed: one could investigate, say, the ship model and specify what inferences model users draw about the target ship from the model, and the pattern that such inferences exhibit, and then, in the inferentialist mode, attempt to use this analysis to explain the relationship that holds between the model and its target.[38] But such a project has yet to be developed in the philosophical literature on scientific representation.

This option, even if it were available, would not help INFERENCE. First, thus understood, the account would pertain to *meta*-semantics, but INFERENCE is an account about the semantics of models, not their meta-semantics. Second, an inferentialist meta-semantics would involve providing an informative account

---

[37] Alternatively, if the analogy relies on the version of inferentialism that demands non-standard semantics (rather than inferential meta-semantics to explain referential semantics), then one could adopt a deflationary account to the semantics of models, which we discuss at the end of this section.

[38] Such a project would be analogous to the prototypical inferentialist analysis of the logical connectives in formal logic. According to such analysis, the meaning of connectives like 'and', 'or', and 'not' is given by their introduction and elimination rules. See Murzi and Steinberger (2017, Sec. 2) for a discussion.

of the semantics of models that goes beyond the identification of 'surface features' or 'platitudes'.

A further issue is that in the context of the philosophy of language it was taken for granted that the semantics of the languages in question are relatively clear (e.g. that 'dog' refers to dogs). Accordingly, the primary thrust of the inferentialist project in the philosophy of language lies in the meta-semantics with the aim of demonstrating that the semantics of sentences can be reconstructed from their inferential role. In contrast, when it comes to scientific models, it is less obvious what their semantics are in the first place. Consider again the ship model. The 'meaning' of this model, what it tells us about the target ship, is not entirely transparent. As such, the question of what is explanatorily prior, the inferences made in model-based science (their use) or the relationship between models and their targets (their semantics), is not the primary focus of the Semantic Question. Even if the inferentialist approach is correct, and we should take the inferential role of scientific models as conceptually basic, we want to know the details of the semantics of models that can be reconstructed from these inferences. By answering the Semantic Question we will get a better grip on what models mean in the first place, and some analysis of this should be given. INFERENCE doesn't help us here.

The second way to support inferentialism about scientific representation is to draw on parallels with deflationary theories of truth (which dovetails with the idea discussed earlier, that the inferentialist could also adopt a non-standard, in this case deflationary, approach to semantics). This is Suárez's own reason for taking the conditions in INFERENCE as conceptually primitive. In philosophical discussions about truth, the central concern is what it means to attach the predicate 'is true' to a sentence. What does it mean, for example, when we say that the sentence 'the dog is on the bed' is true?[39] A seemingly straightforward answer is that when we assign the truth predicate to a sentence, we ascribe to it the property of 'corresponding with the facts'. This is the 'correspondence theory of truth'. Alternatively, we might say that when we assign the truth predicate to a sentence, we are ascribing it the property of being a member of some set of sentences, all of which cohere with one another. This is the 'coherence theory of truth'. Either way, we might think that there is a philosophical job to do in analysing how we are best to understand the property *true*, either in terms of correspondence or coherence.

In contrast, deflationary theories of truth deny that there is a role for any substantial analysis of this kind. This denial can then be developed in different ways, resulting in different kinds of deflationary theories, whose central tenet is

---

[39] For an introduction to different theories of truth, see, for instance, Kirkham (1992).

that 'true' and 'false' do not admit a theoretical elucidation or analysis. Suárez (2015, 44) submits that deflationist accounts of truth should be used to motivate an analogous deflationism about scientific representation: if there is no substantial analysis to be given about truth, there is no substantial analysis to be given about scientific representation.

This second foray into another branch of philosophy faces problems similar to the first. The first, and obvious, worry (a natural analogue of which also applies to the previous approach) is that it assumes that a deflationary theory of truth is appropriate for truth (which is contentious), and that what is good for truth is good for scientific representation (which needs to be motivated).[40] Second, the natural analogue of truth in the context we're currently working in is accurate representation, rather than representation itself. Third, as with the inferentialist approach, for at least some of the theories of truth Suárez discusses, there is a substantial question about what governs the use of the truth predicate *in particular instances*; their 'deflationary' credentials consist in denying that this is the same across all instances (what Suárez calls 'abstract minimalism'), or in denying that they can perform an explanatory role (what Suárez's calls 'use-theory', since uses cannot explain themselves), but informative discussions of particular cases can be given nevertheless.[41] In the context of scientific representation, the question arises again: even if INFERENCE is all that can be said about scientific representation in general, we can still reasonably ask what establishes representational, or inferential, relationships for particular scientific models.[42]

3.4 Reactions and Developments

Following his criticism of INFERENCE quoted at the end of Section 3.2, Contessa claims that when investigating scientific representation 'it is not clear why we should adopt a deflationary attitude *from the start*' (2007, 50, original

---

[40] See Frigg and Nguyen (2020, Sec. 5.3) for detailed discussion of the analogy between truth and scientific representation.

[41] Indeed, as noted by Khalifa, Millson, and Risjord (2022), it's unclear why what Suárez calls 'abstract minimalism' should be thought of as deflationary, rather than substantially pluralist, in the sense that on such a position what it means for a sentence to be true, or for a model to represent its target, seems to require a substantial answer, it's just that this answer may vary across different sentences/models.

[42] Suárez calls these individual model–target relationships 'the means' of representation, and the conditions in INFERENCE 'the constituents' (2015, 46–7) (see also his (2003)). His claim then is that because the former vary across different cases of scientific representation, we're precluded from identifying them with the constituents (2003, Sec. 4.1 in particular). Whilst we disagree on this point, it doesn't matter too much for our current purposes: we would be satisfied with an explication of what 'the means' are, at least in some paradigmatic instances of scientific models. We are told that 'some common means of scientific representation include isomorphism, similarity, instantiation, truth, stipulation' (2004, 768), but little is said beyond this.

emphasis), thereby rejecting the motivations for deflationism. Instead, he provides an 'interpretational account' where an 'interpretation is not just a "symptom" of representation; it is what makes something an epistemic [or scientific] representation of a something else' (2007, 48). So in terms of the question of conceptual priority, it is clear that an 'interpretation' is supposed to play an explanatory role: it's what makes something a scientific representation, and it's what's supposed to explain how scientific models allow for surrogative reasoning.

What is an 'interpretation'? Contessa offers a detailed formal characterisation (2007, 57–62) whose leading idea is as follows: the agent using the scientific model identifies a set of relevant objects in the model, along with a set of properties and relations that these objects instantiate. They do the same in the target. The agent then (a) takes the model to denote the target; (b) takes every identified object in the model to denote exactly one object in the target (and every relevant object in the target to be denoted by exactly one object); and (c) takes every property and relation in the model to denote exactly one property or relation of the same kind in the target (and, again, and every relevant property and relation in the target to be so denoted).[43] A formal rendering of these conditions is what Contessa calls an *analytic interpretation*.[44]

In line with the idea that an interpretation is supposed to provide an explanation of what turns a model into a representation, this provides the following answer to the Semantic Question:

ANALYTIC INTERPRETATION: a model represents its target if and only if a model user adopts an analytic interpretation of the model in terms of the target.[45]

The next question is how an interpretation relates to surrogative reasoning. Since the former is supposed to explain the latter, it is clear that it is *in virtue* of the user adopting an interpretation that a model represents, and licences inferences about, its target. The way that this works is relatively straightforward: if, according to the interpretation, an object in the model denotes an object in the target, then one can infer from the fact that the former exists in the model that the latter exists in the target; similarly, if a relation in the model denotes

---

[43] Two relations are of the same kind if they have the same *arity*. The arity of a property or relation is the number of objects to which it applies in any particular instance, for example, a one-place property like 'is red' applies to single objects when they are red, a two-place relation like 'is taller than' applies to pairs of objects when the first is taller than the second, and a three-place relation like 'is between' applies to triples of objects when the first is between the second and the third.

[44] Contessa's definition includes an additional condition pertaining to functions in the model and target, which we suppress for brevity.

[45] We note here that strictly speaking Contessa doesn't require than an *analytic* interpretation be a necessary condition since 'representations whose standard interpretations are not analytic are at least conceivable' (Contessa 2007, 58), but it is clear that some sort of interpretation is required.

a relation in the target, and there are objects in the model which denote objects in the target, then one can infer from the fact that the model objects are related in the model, that the target objects are related in the target (Contessa 2007, 61). Thus, ANALYTIC INTERPRETATION provides an answer to the Semantic Question that meets the Surrogative Reasoning Condition via the ways in which interpretations provide rules according to which a scientist can infer claims about the target from facts about the model.

Given that ANALYTIC INTERPRETATION works by providing a system of model-to-target inferences, the ways in which it answers the other questions, and handles the other conditions, that have structured our discussion throughout this Element are largely in agreement with the earlier discussion of how INFERENCE does (modulo the fact that we are now offered an explanation of what licences the inferences, in terms of an interpretation), so we won't repeat all of that discussion here. What is important, though, is the Misrepresentation Condition.

Some kinds of misrepresentations are naturally handled by the notion of an interpretation. As an example consider a map of a woody area whose users take the property of *being green* to denote the property *being forested*, and points on the map to denote locations in the terrain. Such users are licenced to infer from the fact that a particular area of the map is green the claim that a particular area of the terrain is forested. But, suppose the map is out of date, and doesn't reflect the fact that there is now some deforested farmland in the target. Then the map is a misrepresentation: it licences a model-to-target inference with a false conclusion. Assuming that there is still some forest in the terrain, which seems to be required by the part of an analytic interpretation that requires that model users take model properties to denote features that the targets *actually have*, ANALYTIC INTERPRETATION can handle this type of misrepresentation. But not all kinds of misrepresentation work this way. Consider a case where the whole of the target terrain is deforested. In which case, how can a model user take the property *is green* to denote the target property *is forested*, given that, by hypothesis, the target doesn't contain the latter property at all? Alternatively, consider a case where the model represents the target as including an object that isn't there; how can the interpretation specify that an object in the model denotes a non-existing object in the target? As Shech (2015) points out, ANALYTIC INTERPRETATION seems to have difficulties accommodating these kinds of misrepresentation.[46]

Conversely, and like the resemblance accounts discussed in the previous section, there are cases of accurate scientific representation that don't, at least in any obvious way, turn on inferences generated according to an analytic interpretation. As a case in point consider the Ising model of phase transitions,

---

[46] For further criticisms, see Bolinksa (2013) and Ducheyne (2012).

a phenomenon that we observe, for instance, when water starts boiling or a metal becomes magnetic. The model involves taking the thermodynamic limit, and as such the model contains an infinite number of lattice sites.[47] The fact that the model contains an infinite number of objects precludes it being the case that each object in the model denote one and only one object in the target and that every object in the former be so denoted. This is because target systems like a pot of water, or a metal bar, consist of a finite number of molecules, and an infinite set cannot be put into a one-to-one correspondence with a finite set. But nevertheless, the model provides us with lots of (true) information about how phase transitions take place, and indeed forms the basis of our conceptual understanding of them.

More generally, given that an analytic interpretation requires the exact match between model and target that is also required by an isomorphism, ANALYTIC INTERPRETATION fails to accommodate practices where scientists exploit model–target distortions in much the same way that resemblance accounts based on the notion of an isomorphism failed.[48] At best, it is an open question whether these sorts of cases can be reconstructed in such a way that they fit the mould of an analytic interpretation.

A solution to the former worry (about denoting non-existing objects, properties, and relations) emerges from Díez's (2020) 'ensemble-plus-standing-for' account, which relies on the idea that a model user conceptualises both the model and the target as an 'ensemble', that is, a collection of objects and properties and relations, and then associates the ensemble corresponding to the model with the ensemble corresponding to the target in a manner similar to Contessa's interpretation. However, Díez also distinguishes between 'performance conditions' and 'adequacy conditions' for scientific representation, which answer our Semantic Question and Accuracy Question, respectively. Intuitively, performance conditions say what an agent has to do in order to use a model to represent a target in the first instance; adequacy conditions specify what the agent has to do to represent the target accurately. Díez's (2020, 140) performance conditions for an agent to use a model to represent a target are that:

> ENSEMBLE PLUS STANDING FOR: a model represents its target if and only if: (1) both the model and the target are ensembles; (2) the agent intends to use the

---

[47] See Baxter (1982) for a technical discussion of the model, and Butterfield (2011) for a philosophical one.

[48] The difference between the two approaches lies in the fact that ANALYTIC INTERPRETATION treats models and targets as fully fledged entities, rather than extensional structures. But in terms of how models and their targets can be mismatched with respect to their structural features, and yet the former can still yield true information about the latter, the same objection applies to both.

model to represent the target; (3) the agent intends that the (contextually relevant) constituents of the model stand for (contextually relevant) constituents of the target; and (4) all (contextually relevant) constituents of the target are stood for by a (contextually relevant) constituent of the model.

The success conditions add to the performance conditions that (i) the postulated intended target entities exist; (ii) that the stood for entities in the target behave as their corresponding entities in the model; and (iii) that the agent's purposes are served by using the model to represent the target in virtue of (ii) (2020, 142). In this way Díez allows for cases where the model user conceptualises the target ensemble in a way that involves it consisting of objects, properties, and relations, which needn't actually be there, thus accommodating the kind of misrepresentations that challenged Contessa's account because Díez is explicit that intentions – invoked in condition (3) – can fail to bear out (2020, 141).

However, again, this account doesn't accommodate cases where scientists exploit model–target mismatches in order to generate true conclusions about the features of the latter from the features of the former. Returning to the Ising model, it is at best unclear how condition (3) can be met in a case where a model user exploits the fact that there are an infinite number of objects in the model to reason about the finite number of objects in the target. And again, this isn't a problem particular to the Ising model: ENSEMBLE PLUS STANDING FOR has difficulty accommodating any case where structural mismatches underpin the inferences drawn by the model user. So the question arises again: how are we to understand surrogative inferences of this kind? We provide our preferred answer to this question in Section 4.

In this section, we have seen how one could draw on discussions from elsewhere in philosophy to motivate alternative ways of answering the Semantic Question: rather than expecting explanatory conditions on scientific representation, we could take the fact that they allow for inferences about their targets as conceptually primitive. However, we have also argued that this way of motivating such an approach faces a number of questions; in particular, it suggests a framework for thinking about model–target relationships that still requires specifying what they are (even if they are explained by, rather than explain, the inferences that are drawn, and even if they differ across different instances of scientific representation) and thus the deflationary attitude expressed by INFERENCE (i.e. the idea that there is nothing more to be said) remains unmotivated. We have also seen that alternative accounts that are, at least in part, motivated by the dissatisfaction with this aspect of INFERENCE face similar objections to those facing the resemblance accounts discussed in the previous section: the cost of introducing explanatory conditions on scientific representation is paid by requiring that model-to-target inferences are underpinned by tight structural associations between models and their targets, where these associations are, at the very least, not met in any clear way by various instances of accurate scientific representation.

4 The DEKI Account

4.1 Introduction

We now turn to our own preferred account of representation, which we call the DEKI account to indicate its main components: denotation, exemplification, keying up, and imputation. The account is explicitly designed to capture the sort of reasoning that the previous accounts have difficulty accommodating: cases where model users exploit model–target mismatches during their surrogative reasoning. We start our discussion with our example of the ship. As we will see, working through the example in detail furnishes all the essential components of the account (Section 4.2). This sets the discussion on the right path, and the next steps are to refine and develop the notions thus introduced. We elaborate the distinction between being a representation of something and being a something-representation (Section 4.3), introduce exemplification (Section 4.4), and discuss the notion of a key (Section 4.5). Combining these elements yields the DEKI account (Section 4.6). All of this is illustrated with the ship model, which is a concrete model. But, as discussed previously, not all models are like this; some are held in our heads, not our hands. The next step is to show how the DEKI account applies to non-concrete models (Section 4.7) and to illustrate with our second example, the bridge jump, how the account works in such a case, and what key is used to connect the model to the target (Section 4.8). We end with a few afterthoughts (Section 4.9).

4.2 Using a Ship Model

Return to the example in Section 1.1. You're a shipbuilder tasked with redesigning SS *Monterey*'s propulsion system so that it has enough power to sail at a certain speed. To this end, you need to know what resistance the ship faces at that speed. You don't have data of past performance available, and theoretical calculations are out of reach. So you decide to use a 1:100 scale model of the ship and study its behaviour in a tow tank.[49] The thinking is that the model represents the real ship, and so you expect be able to gain information about the

---

[49] Despite their importance in historical, and contemporary, science, scale models like this have received relatively little attention from philosophers of science working on models and scientific representation. Important exceptions include Weisberg (2013), Sterrett (2017a, 2017b, 2020), and Pincock (2022).

ship from the model. In principle this is correct, but the model doesn't yield easily. How should you extrapolate from the resistance faced by the model to the resistance faced by the ship? One might think this is straightforward: simply multiply the resistance of the model with the scale to get the resistance of the ship. On this view, if we measure resistance R on the 1:100 model, the ship's resistance would be 100 R. This is both unjustified and wrong. It's unjustified because one cannot assume without further argument that quantities extrapolate with the model's scale. Many don't. As a simple example consider a 1:s scale model of a cube (where s is the scale factor on the length of the cube's side). Suppose you measure the volume of the model cube and multiply it by s to find the volume of the real cube. Your result is wrong. It's easy to see that volume scales with s^3: the volume of the original cube is s^3 times the volume of the model cube. Of course, this doesn't preclude that there are some quantities for which it is correct to multiply them with the scale factor, the cube's length is an obvious one in this case. However, as we will see in this section, resistance is not like length, and so the simple extrapolation is also wrong.

British engineer William Froude systematically studied the relation between ships and their models in the nineteenth century, and his approach became the foundation of the modern theory of ship design. In the remainder of this section we introduce the main ideas of Froude's approach.[50]

The resistance a ship experiences when moving through water is due to a number of different factors. Froude realised that there are two principal components to ship resistance: frictional resistance and wave-making resistance. *Frictional resistance* (R_F) is the resistance created when an object is dragged through water. It can be compared to the friction an object experiences when it's dragged along the floor. *Wave-making resistance* (R_W) is the resistance an object experiences due to making waves. Making waves effectively means displacing water, which requires a force that the ship experiences as resistance. The complete resistance of the ship (R_C) is the sum of these components:

$$R_C = R_F + R_W \tag{4}$$

Froude's crucial innovation was to devise methods to determine these components. We begin with frictional resistance. R_F depends on a number of factors, including the roughness of the surface, the size of the object under water, and the speed at which it moves. Froude performed a number of experiments at a tank in Torquay that involved dragging a series of planks through water and measuring

---

[50] Our presentation of Froude's method follows Carlton (2007, Ch. 12). We use Froude's method because it has all the features that matter for our philosophical discussion while still being readily accessible to non-experts. The DEKI account we develop in this section has no problem accommodating more modern approaches as discussed in, for instance, Molland (2008).

the resistance they experienced. The planks were different lengths, with different surface finishes such as shellac varnish, paraffin wax, tin foil, and bare wood. From these experiments Froude derived an empirical formula for the calculation of frictional resistance:

$$R_F = f\, S\, V^n, \tag{5}$$

where n is an empirical constant with the value 1.825; f is a coefficient that depends on the roughness of the surface and the length of the ship; S is the wetted surface area of the ship; and V is the velocity at which it moves.

Things get more involved when it comes to determining the wave-making resistance, and this is where the model enters the scene. In what follows, we use the superscript 'M' to indicate that a quantity pertains to the model, and superscript 'T' if it pertains to the target, that is, the real ship. The initial task is to determine the model's wave-making resistance, R_W^M. To this end, you first drag the model through the tank and measure its complete resistance R_C^M. You then use Equation (5) to calculate the frictional resistance R_F^M of the model. You then, finally, turn to Equation (4) and calculate the wave-making resistance of the model from these two quantities: $R_W^M = R_C^M - R_F^M$.

The next task is to extrapolate from R_W^M to R_W^T, the wave-making resistance of the real ship. This is done in two steps. The first concerns the relevant velocities. The wave-making resistance depends on how fast an object moves, which can be made explicit by noting that wave-making resistance is a function of velocity: $R_W(V)$. So if you want infer from R_W^M to R_W^T, you have to decide at what velocities these quantities should be considered. One might think there's not much of a question here: surely it's obvious that you want them at the same velocity. It's one of Froude's great insights that this is wrong. As noted, the wave-making resistance a ship experiences is the result of pushing water away, and how much resistance that is depends on *how* that happens. As it turns out, model–target comparisons are best made when they produce the same wave pattern, and Froude noted that this does *not* happen when they move at the same speed. Rather, a ship and its geometrically similar scale model produce the same wave pattern if their velocities V^T and V^M are related by:

$$\frac{V^M}{V^T} = \sqrt{\frac{L^M}{L^T}}, \tag{6}$$

where L^T and L^M are, respectively, the ship's and the model's lengths. Velocities that are related by this equation are *corresponding velocities*. As a result, if you want to study what wave-making resistance a ship travelling at V^T experiences, you have to study the wave-making resistance of model travelling at $V^M = V^T \sqrt{L^M/L^T}$.

The second step is to calculate $R^T_W(V^T)$ from $R^M_W(V^M)$. One might think that once one compares resistances at corresponding velocities, the two resistances are related to each other by the model's scale; i.e. $R^T_W(V^T) = 100 \, R^M_W(V^M)$. But that's, again, wrong. In fact, Froude noted through observations that resistances at corresponding velocities are proportional to the quotient of the objects' displacements (i.e. the quantity of water that they 'push away' when they are floating):

$$R^T_W(V^T) = \frac{\Delta^T}{\Delta^M} R^M_W(V^M),$$ (7)

where Δ^T and Δ^M are, respectively, the displacements of the target and the model.

Taking all this together, the complete resistance $R^T_C(V^T)$ of the ship travelling at target velocity V^T is:

$$R^T_C(V^T) = R^T_F(V^T) + R^T_W(V^T),$$ (8)

where the frictional resistance $R^T_F(V^T)$ is calculated with Equation (5) and $R^T_W(V^T)$ is given by the behaviour in the model, combined with Equation (7). This equation answers your initial question because it tells you what resistance the engine must be able to overcome for the ship to travel at V^T.

Let us now step back and reflect on what we have learned about representation from this case. Four points stand out, and versions of all of them will play a crucial role in the formulation of the DEKI account.

First, a ship-shaped block of wood has been used as the model that represents the real ship. In his discussion of pictorial representation Goodman notes that '[a] picture that represents . . . an object refers to and, more particularly, *denotes* it. Denotation is the core of representation' (1976, 5, original emphasis). What goes for pictures equally holds of ship models. What turns the block of wood into a representation of a ship is that it denotes the ship. Moreover, parts of the model denote parts of the target: the tip of model denotes the bow of the ship, and so on.

Proper names are paradigmatic examples of denoting symbols. For example, speakers can use the name 'Mahatma Gandhi' to denote a particular individual, namely Gandhi. But denotation is not limited to linguistic symbols; pictures, graphs, charts, diagrams, maps, and drawings also represent their subjects by denoting them. Denotation has been extensively discussed in the philosophy of language, and we cannot revisit these discussions here. There is, however, one point that deserves mention because it bears directly on one of our conditions of success. Denotation is a dyadic relation that obtains between certain symbols and certain objects. The relation can hold only if both relata exist, implying that something that does not exist cannot be denoted. Targetless models lack a target and so whatever representational function they perform cannot be analysed in terms of denotation: targetless models don't denote.[51]

Second, you made sure that the model has the same shape as the real ship and then you focussed on the model's resistance, measured experimentally when dragging the ship through the towing tank. The ship has a large number of other features like colour, temperature, and production history, but the experiment disregards these completely. The investigation singles out a small set of highly specific model features as relevant, and it focusses solely on these while disregarding all the others.

Third, model features are not carried over one-to-one to the target; there is no simple feature sharing. And what is more, model features cannot be turned into target features simply by rescaling them with the model's scale factor. Model features are connected to target features through complex relations like the one in Equation (7). We say that this equation serves as the *key* that translates model features into target features. Keys are often complex and result from a mixture of background theory and empirical observation. You can reason surrogatively with a model only once you know the model's key.[52]

Fourth, you impute the output of the key – in our example the left-hand-side of Equation (7) – to the target by saying that the ship is subject to wave-making resistance $R_W^T(V^T)$. It is crucial to note that this is a feature ascription, and it can fail. You may have made an error somewhere, or Froude's account could simply be wrong. A model represents the target as having a certain feature; it does not also say that the attribution is true – it may or may not be. Whether the attribution is true, and what reasons there are to regard it so, is a matter that has to be settled elsewhere. In that sense, models are no different from other representations. A sentence may assert that Italy is north of Germany, and a painting may represent a petite ballerina as being tall. It's not part of the *content* of either the sentence or the painting that these ascriptions are true.

---

[51] For a discussion of models and denotation, see Salis, Frigg and Nguyen (2020) and Salis and Frigg (2020).

[52] As discussed in Section 2, resemblance accounts of scientific representation have difficulty accounting for this style of reasoning: appealing to resemblance to account for the model–target relation in this instance is either wrong (because the ship and the model don't resemble one another in any obvious sense) or incomplete (because it fails to account for how model–target mismatches are crucial in the reasoning process), and the label 'resemblance' is attached after the fact, once this reasoning has been explicated.

These are important and, we submit, valid points. Our task now is to analyse, develop, and generalise them.

4.3 Two Kinds of Representations

We said that we used a ship-shaped object as a model. That's correct, but what turns it into a ship model? With a nod to proverbial ducks we might reply: if it looks like a ship, swims like a ship, and makes waves like a ship, then it probably is a ship (just a smaller one). While that's not too far off the mark in the case we discussed in the last section, in general things are bit more involved. You might not have had the budget to produce an exact scale model and so you might have decided to conduct an investigation by dragging a barrel through water. In that context, the barrel is ship model even though it doesn't look like a ship. In other cases, the discrepancy between the object that serves a model and what it models is even larger. Economists Phillips and Newlyn used a hydraulic machine consisting of water pipes, reservoirs, levers, and pumps as a model of an economy.[53] But an economy doesn't look like such a machine by any standard. Likewise, plasticine sausages are used as models for proteins; balls connected by sticks function as models for molecules; and electrical circuits are studied as models for brain function. We encounter the same situation when looking at non-concrete models (to which we turn in Section 4.7). Imaginary chequer boards are used as models for social segregation; a swarm of imaginary billiard balls serves as a model of a gas; and the mathematical structure of a Hilbert space is employed to model a hydrogen atom. None of these objects look anything like the things they are models for, and, more generally, there is nothing in their intrinsic features that would make them models, let alone models for something in particular. Most hydraulic machines are just a piece of plumbing; they aren't models for anything.

The fact that some objects are models must therefore be rooted in something other than the idea that they somehow look like the objects that they are models for. We submit that this something else is an *interpretation*. We turn a hydraulic machine into a model for an economy by interpreting some of its features in terms of economic features: we interpret the flow of water as the flow of money, the setting of a valve as the tax rate, the level of water in certain reservoir as the reserves of the bank, and so on. The same mechanism is at work in the barrel model, although the presence of an interpretation is admittedly less obvious in this case. Per se, the model is just a collection of wooden staves bound together by metal hoops. It takes an interpretation to turn it into a ship model, whereby

---

[53] For a discussion of this case, see Morgan (2012, Ch. 5), Frigg and Nguyen (2020, Sec. 8.1), and references therein.

the interpretation in effect correlates parts of the model with ship features (the front of the barrel is interpreted as a bow, the back as a stern, the submerged part as the wetted area of a hull, and so on).

'Interpretation' is used with many different meanings, and it is therefore important that we define what we mean. Let X be the object that serves as a model – the barrel, the hydraulic machine, the electric circuit, and so on – and let $X = \{X_1, \ldots, X_n\}$ be a set of features pertaining to X. Next, let Z be the domain we are interested in – shipping, the economy, brain function, and so on – and let $Z = \{Z_1, \ldots, Z_m\}$ be a set of features pertaining to Z. An interpretation I is a one-to-one mapping from X to Z. In effect, the interpretation correlates features pertaining to X with features pertaining to Z, for instance by saying that the front of the barrel is a ship's bow or that the tank in the middle of the hydraulic machine is a central bank. When features are quantitative, the interpretation also contains a scaling factor, which correlates quantities, for instance by saying that one litre of water corresponds to a million of the model-currency. It is not mandated that X and Z be distinct: in cases where the model and target are the same kind of object, such as your original ship model, they are the same and the interpretation functions as identity; mapping the bow of the model to a ship's bow, and so on.

We now define a model M as the ordered pair consisting of the object X and the interpretation I: $M = (X, I)$. This captures how a ship shaped object (or barrel) and a hydraulic machine are turned into models.

An immediate consequence of this definition is that models need not have a target. This seems counterintuitive at first, but it's actually an advantage of the definition. As we have seen in Section 1.3 (e.g. by considering a variant of the ship example where you're tasked with designing a new ship which is ultimately never built), there are targetless models, and an account of representation has to explain how this can be. Our account does this naturally because models ipso facto need not have targets at all.

However, a residual tension remains. There is pervasive intuition that a ship model is a representation of a ship, even if there is no real-world target system (as in the case where you design a new ship). But this is ruled out by the posit, at the end of the previous section, that denotation is the core of representation. In order for x to be a representation of y, x has to denote y, and therefore y has to exist. This doesn't seem to sit well with the notion that models need not have a target. This tension is dissolved by introducing the distinction between a Z-representation and representation of a Z due to Goodman (1976) and Elgin (1983). Let Z be any subject that might be shown in a representation: a horse, a dragon, Europe, Zeus, Mahatma Gandhi, and so on (we use the symbol 'Z', which we have already used in our definition of an interpretation, for reasons that will become clear soon). The crucial insight, in Goodman's words, is that 'representation' is ambiguous:

> What tends to mislead us is that such locutions as 'picture of' and 'represents' have the appearance of mannerly two-place predicates and can sometimes be so interpreted. But 'picture of Pickwick' and 'represents a unicorn' are better considered unbreakable one-place predicates, or class terms, like 'desk' and 'table'. . . . Saying that a picture represents a soandso is thus highly ambiguous between saying that the picture denotes and saying what kind of picture it is. Some confusion can be avoided if in the latter case we speak rather of a 'Pickwick-representing-picture' or a 'unicorn-representing-picture' . . . or, for short, of a 'Pickwick-picture' or 'unicorn-picture' . . . Obviously a picture cannot, barring equivocation, both represent Pickwick and represent nothing. But a picture maybe of a certain kind – be a Pickwick-picture . . . – without representing anything (Goodman 1976, 21–2).

This diagnosis can be captured by distinguishing between a Z-representation and a representation of a Z. Something is a Z-representation if it portrays Z. Phidias' Statue of Zeus at Olympia was a Greek-God-representation, and part of Raphael's *Saint George and the Dragon* is a dragon-representation. Something is a representation *of* a Z if it denotes a Z. George Stubbs' *Whistlejacket* is a representation of a horse because it denotes a horse; and Philip Jackson's statue in the north-west corner of London's Parliament Square is representation of Mahatma Gandhi because it denotes him.

Being a Z-representation and being a representation of a Z are independent notions. Phidias' statue and Raphael's painting are Z-representations but not representations of a Z, because neither Zeus nor dragons exist, and, as noted earlier, only things that exist can be denoted. One might now think that independence fails the other way because Stubbs' painting and Jackson's statue not only denote a horse and Gandhi but also are, respectively, horse-representations and Gandhi-representations. While correct for the specific examples, the point doesn't generalise. The word 'Europe' is representation of Europe (because it denotes Europe) but it is not a Europe-representation (it doesn't show or display Europe, as, for instance, a map of Europe would), and Edvard Munch's *The Scream* is a man-representation but it's arguably not a representation of man, but a representation of the anxiety inherent in the human condition. Hence, as Goodman notes, the kind a representation is independent of what is denotes: 'the denotation of a picture no more determines its kind than the kind of picture determines the denotation. Not every man-picture represents a man, and conversely not every picture that represents a man is a man-picture' (1976, 26).

There is nothing special about pictures and statues, and the distinction between Z-representations and representations of a Z equally applies to scientific models. Take Maxwell's well-known ether model. In our idiom this is an ether-representation because the model portrays the ether, but it is not a model of the ether because there is no ether. Contrast this with Bohr's hydrogen model. That model is both a hydrogen-representation (because it portrays hydrogen) and representation of hydrogen (because it denotes hydrogen). A verbal description of the properties of the Higgs boson is a representation of the Higgs boson (it denotes the Higgs boson) but it's not a Higgs-boson-representation because it doesn't portray a Higgs boson. Cases of models that work like *The Scream* are arguably less common, but they nevertheless occur. The Chicago Police Department is now using epidemiological models to first model and then fight crime. So here we have model that is an epidemic-representation, which is used as a representation of crime.[54]

We have so far spoken somewhat loosely and said that a Z-representation portrays a Z. In the visual arts, this notion is most commonly unpacked by appealing to a spectator's perceptual experience, and the posit that what kind of picture a picture is depends on the perceptual experience we have when looking at it. A picture is, say, a horse-representation because we experience seeing a horse when observing it. What the relevant experience consists in is a matter of controversy, with influential suggestions being that pictures give us the *illusion* of viewing what we see in the representation (associated with Gombrich), or that we have the perceptual skill of *seeing-in* (associated with Wollheim).[55]

Whatever merit these proposals have in the context of visual representation, they don't explain how models are Z-representations because models don't, generally speaking at least, work with spectators' perceptual experience. A hydraulic machine doesn't provide an onlooker the perceptual experience of an economy, and electric circuits don't look like brain functions. Models require a different account. The good news is that, without realising it, we've already developed this when we introduced the notion of an interpretation (which was the reason for keeping the symbol 'Z'): what makes an object a Z-representation is that it's interpreted in terms of Z. A barrel becomes a ship-representation if it is interpreted in terms of naval terms, and a hydraulic machine becomes an economy-representation if its parts are interpreted in economic terms. The relevant sense of 'interpretation' is the one introduced previously, and so we have reached the conclusion that models, understood as ordered pairs involving an object and an interpretation, are ipso facto

---

[54] For discussion of the approach, see Slutkin (2013) and Wiley *et al.* (2016).
[55] For discussion of the various options, see Kulvicki (2006).

Z-representations! Some models then *also* denote a target, which makes them representations of the target. This happens with the ship model when you use it to redesign the ship. But as the example of designing a ship that is never constructed demonstrates, this is not built into the definition of a model.

4.4 Exemplification

When discussing the ship in Section 4.2 we noted that the investigation singles out a few features as relevant while disregarding all other aspects of the model: you focus on the model's resistance not its colour. How are we to understand this 'selectiveness' from the point of view of an account of representation? Here's where the notion of exemplification enters the scene.

Set models aside for a moment and imagine you want to redecorate your living room. You go to a paint shop to decide on the colour of the chimney breast. In the shop you look at colour swatches displaying your options. Eventually you look at swatch of royal blue and decide that that's the colour for you. The swatch represents royal blue, and it does so by *exemplifying* it. An item exemplifies a feature if it at once instantiates the feature and refers to it (Goodman 1976, 53; Elgin 1983, 71). An item that exemplifies a feature is an *exemplar*. The conjunction of instantiation and reference is crucial. An item that doesn't instantiate a feature cannot exemplify it. The word 'royal blue' refers to royal blue, but it does not exemplify it because it doesn't instantiate that colour. Instantiation is necessary for exemplification. But it's not sufficient. Exemplification is selective: the swatch is also rectangular, printed on acid-free paper, weighs 10 grams, and has been in the shop since last year, but these features are not exemplified by the swatch.

What marks the difference between instantiated features that are 'merely instantiated' and those that are exemplified? The difference isn't grounded in the features themselves, nor is it dictated by the object itself. Neither physics nor metaphysics can determine which of an object's features are exemplified. The determining factor is context. If we bring the swatch to geometry class, it can exemplify rectangularity, and if used by a paper merchant, it can exemplify being acid-free. Exemplification, unlike instantiation, is determined by context. One might now be tempted to ask for a rigorous definition of a context. We doubt that such definition is available. But we also think that not much is lost by not having a definition. For the purpose of an account of representation, it is sufficient to think of a context as a certain set of problems and questions that are addressed by a group of scientists using certain methodologies while being committed to certain norms (and, possibly, values).

One aspect of the context is crucial: from an exemplar we can learn about its exemplified features, and this requires that the context offers epistemic access to exemplified features. A colour swatch that's too small to see with the naked eye does not, in the context of a paint shop, exemplify royal blue, even if the colour could, in principle be seen under a light microscope. An exemplar is therefore not merely an instance of a feature but a *telling instance* (Elgin 2010, 5).

We can now answer the question we asked at the beginning of this section: the relevant sense of 'selectiveness' that we encounter in models is exemplification. The ship model exemplifies its resistance because it both instantiates resistance and refers to it in the same way in which a colour swatch has a certain colour and refers to that colour. Resistance is epistemically accessible in the context of a towing tank; it's not epistemically accessible if the model sits in the window of travel agent. So the ship exemplifies resistance in the former but not the latter context.

One last tweak is required to make this a fully general account. Shifting focus from the ship example to the example of the hydraulic machine reveals an issue. Exemplification requires instantiation. A hydraulic system instantiates hydraulic features like having a flow of two litres per minute through a certain pipe; it doesn't instantiate economic features like two million of the model-currency being received by the treasury. So it seems that our notion of exemplification forces the unfortunate consequence on us that the economy-representation can only exemplify hydraulic features. We want a model to allow us to learn about the domain that it represents, and we want exemplification to be the instrument with which we learn. The mismatch of the features that are exemplified and the features that we aim to learn about poses a problem.

Fortunately, there's an easy fix. Nothing in what we want to do with models depends on features being instantiated *literally*. In the previous section we introduced the notion of an interpretation, which establishes a one-to-one correspondence of features of the model-object X with features of the domain that the object represents. This correspondence can be exploited to introduce the notion of *instantiation-under-interpretation-I*, or, for short, *I-instantiation*. The idea is simple. The interpretation correlates X-features with Z-features, and so we can say that the model I-instantiates a certain Z-feature iff it instantiates the corresponding X-feature. Or more precisely: $M = (X, I)$ I-instantiates a Z-feature Z_j iff X instantiates an X-feature X_i such that X_i is mapped onto Z_j under I (and if the features are quantitative, the associated values of the features are also correlated by I). Nothing in our notion of exemplification depends on features being instantiated in a metaphysically robust sense; they can just as well be I-instantiated. This allows us to introduce the notion of

I-exemplification, which is exactly like exemplification except that features are *I*-instantiated rather than instantiated.

4.5 Keys and Imputation

In our example with the ship, the model does not exemplify the feature that was ultimately imputed to the target system. You use Equation (7) to convert a model feature (the wave-making resistance of the model) into a significantly different target feature (the wave-making resistance of the ship). This is not an idiosyncrasy of the example. On the contrary, it is a typical feature of most scientific models. Ever since scientific modelling has attracted the attention of philosophers, they have emphasised that models simplify and distort.[56] This is also illustrated in our other example, the bridge jump, where there is no air resistance in the model, the car is a perfect sphere moving on a perfect geometrical plane, and all forces other than the earth's gravity are eliminated. This means that models typically don't exemplify exactly the features that we take the target to have, and that model features typically should not be transferred to a target *unaltered*.[57]

An account of representation has to take this into account and create a systematic space for this 'conversion' of model features into target features. In our account, this is done by what we call the *key*, which is in effect a systematic 'translation manual' that tells us how to convert exemplified model features into target features. More precisely, consider the set $Z = \{Z_1, \ldots, Z_m\}$ of features that are exemplified by the model. A key K associates the set Z with a set $Y = \{Y_1, \ldots, Y_l\}$ of features that are candidates for imputation to the target system. Using algebraic notation somewhat loosely, we can write: $K(\{Z_1, \ldots, Z_m\}) = \{Y_1, \ldots, Y_l\}$. In the ship example, both Z and Y have only one feature – respectively, the model's wave-making resistance Z_1 and the to-be-imputed ship wave-making resistance Y_1 – and the key connecting the two is given by Equation (7).

At this point one is tempted to ask: what are the keys that are used in various representations? This is a good and important question, but not one that a philosophical account can answer. K is, as it were, a blank to be filled. We have established that there is a blank; to fill it is a task for model users. The key

---

[56] Classic discussions include the contributions to Morgan and Morrison (1999). See also Elgin (2017) and Potochnik (2017) for more recent discussions.

[57] The same can also be said for the examples in Section 2 that were not models: both the tube map and the fractal images exemplify their colour properties. But they don't represent their targets as having such colours. The tube map is accompanied by a key associating distinct colours with distinct tube lines (e.g. red is mapped to the Central line), and the keys accompanying the fractal images associate colour with the speed at which a function diverges.

associated with a particular model depends on many factors: the scientific discipline, the context, the aims and purposes for which the model is used, the theoretical backdrop against which the model operates, and so on. In our example, nautical engineers figured out how to convert features of the model ship into features of the target. In Section 4.8, we will discuss how mechanical models like the bridge jump model relate to their targets. Similar considerations will have to be made in every individual case and there is no one-size-fits-all solution. The important point is that model users are aware that there is a blank to be filled, and that they don't naïvely assume that reality is just like the model.

The last step is to impute at least one of the features in Y to the target. We do this in the example by saying that the real ship has feature Y_1. So imputation can be understood as the model user ascribing a feature to the target. This step looks simple compared to the others, and in sense it is. Nevertheless, there is an important pitfall. The user ascribes features to the target T, but this does not imply that T actually has these features. A representation is *accurate* if T indeed possesses the features that M imputes to it.[58] But that M is accurate is not part of it being a representation. Indeed M may represent T as having all the features in Y, and yet T could turn out not to have a single one. This ensures that the account meets the Misrepresentation Condition.

4.6 Putting the Pieces Together: DEKI

We now have all elements of our account on the table and can put them together to form what we call the DEKI account of scientific representation:

> DEKI: Let $M = (X, I)$ be a model, where X is an object and I is an interpretation. M represents T if and only if the following four conditions are satisfied:
>
> (i) M denotes T (and in some cases parts of M denote parts of T).
> (ii) M exemplifies Z-features $Z = \{Z_1, \ldots, Z_m\}$.
> (iii) M comes with a key K which associates the features in Z with another set $Y = \{Y_1, \ldots, Y_l\}$ of features: $K(\{Z_1, \ldots, Z_m\}) = \{Y_1, \ldots, Y_l\}$.
> (iv) M imputes at least one feature in Y to T.

As noted previously, we call this the DEKI account of representation to highlight its defining features: denotation, exemplification, keying-up, and imputation.[59] The account's structure can be schematically visualised as shown in Figure 6.

---

[58] Of course, models themselves don't literally attribute features to targets; we use the locution as a shorthand for the idea that the model user does so.

[59] DEKI is a close cousin of the notion of representation-as developed by Goodman (1976) and Elgin (2010). In Frigg and Nguyen (2017), we discuss in detail how the DEKI conditions relate to their approach. Hughes (1997) develops another way of thinking about scientific representation,

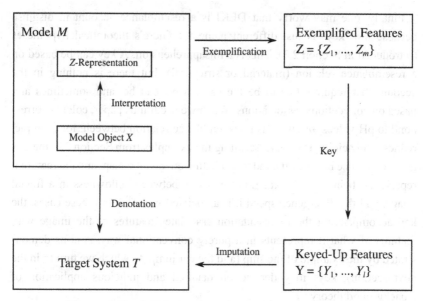

Figure 6 The DEKI account of scientific representation

It's worth adding a few qualifications. First, the account doesn't presuppose that the features in Z and Y are monadic or independent of one another. They can include relational features, and they can enter into all kinds of relations with each other. Second, unlike the interpretation, which establishes a one-to-one correspondence, the key need not be one-to-one. In fact, it needn't have the structure of a function from Z to Y. Any operation that transforms the members of Z into other features can, in principle, serve as a key. Third, as previously noted, 'scientific model' is not a synonym for 'scientific representation'. By definition, all models are Z-representations, but not all models are representations of a target. Some models are targetless models and hence don't denote anything. Fourth, what we previously said about the key holds true about all elements of DEKI. The conditions stated previously provide the general form of an account of representation and they need to be concretised in every particular instance: denotation has to be established, the interpretation has to be specified, context has to specify which features are exemplified, and so on. Depending on what kind of model we are dealing with, these things will be done in different ways. We take this to be an advantage of the account because models are diverse and an account that is too specific will invariably either limit this diversity or fail to cover some of the cases.

---

which is also inspired by Goodman and Elgin's work; see Frigg and Nguyen (2020, Sec. 7.2) for a discussion.

Finally, one may worry that DEKI is a resemblance account in disguise because the 'key' is just different name for Giere's theoretical hypotheses (introduced in Section 2.3). This is a misapprehension. A key can be based on a resemblance relation (material or structural), but there is nothing in the account that requires this to be the case. Keys can be, and sometimes are, based on conventional associations. When we use litmus paper, colours correspond to pH values, but there is no resemblance relation between, for example, redness and acidity. Likewise, recalling the examples from Section 2.4, there is no resemblance between the redness of a line on the tube map, and the feature it represents (being the Central Line), nor one between yellowness in a fractal image, and the divergence speed it is associated with.[60] In all these cases, the key accompanying the representation associates features of the image with features of what it represents in a purely conventional way. And as demonstrated by the example of the ship (and the car jump, to which we turn to in the next section), keys may depend on detailed and judicious application of a background theory.

We now discuss how DEKI deals with the questions and conditions we introduced in Section 1. The Semantic Question asks: in virtue of what does a model represent its target? Our answer: in virtue of meeting the DEKI conditions. This question has four conditions associated with it that a successful answer must meet. The first is the Directionality Condition. DEKI requires that M denotes T and since denotation is an asymmetrical relation, DEKI accounts for the directionality of representation. While this suffices to meet the condition, DEKI has further in-built asymmetries: M exemplifies various features, T doesn't (or at least doesn't necessarily); M comes with a key associating these features with those to be imputed, T doesn't; and the model imputes those features to T without imputing anything to M.

With respect to the Surrogative Reasoning Condition, DEKI offers a clear account of how we generate hypotheses about the target from the model: we figure out which features are exemplified, convert these into other features using the key, and impute these to the target. This is how surrogative reasoning works. The Misrepresentation Condition is met by construction. As we noted previously, that M imputes features to T neither implies nor presupposes that T actually has these features. Additionally, misrepresentation can also happen via the denotation condition because denotation can fail in various ways: a representation can purportedly denote a target that does not exist, or it can denote the wrong target. Finally, the Targetless Models Condition is also met by construction. In DEKI, models are Z-representations and because Z-representations don't have to be

---

representations of anything – that is, don't have to denote anything – there is no problem in having non-denoting *Z*-representations, and this is precisely what targetless models are.

Next is the Accuracy Question: what turns something into an accurate representation of something else? As we have seen in the previous section, a model is accurate if *T* possesses the features that *M* imputes to it. This meets the Gradation Condition because no restriction is imposed on allowable features. A key can output a precise feature (like 'the resistance of the ship when travelling at 10 knots is 11.657×10^4 N') or an imprecise feature (like 'the resistance of the ship when travelling at 10 knots is between 9×10^4 N and 12×10^4 N'). And for a fixed set of features imputed, *T* can actually instantiate any number of them. Thus, the more imputed features *T* instantiates the more accurate the representation is (with respect to those features). Both of these aspects of the account make room for varying degrees of imprecision and inaccuracy, and thus accommodate the Gradition Condition. The Contextuality Condition is met trivially. As we have already seen, context is built into the notion of exemplification. And when constructing their key, model users are free to choose which features they want a model to provide and how precisely specified these features are, and model users can obviously take their context into account when the make these choices. As a result, how accurate a model is will depend on the context in which the model is used.

Finally, the Model Question: what objects are these models? For the models we have discussed so far, models like the ship and the hydraulic machine, the answer is straightforward, namely that they are concrete objects. The conditions that pertain to this question – the Model-Truth Condition, the Model-Epistemology Condition, and the Model-Identity Condition – raise many interesting issues, but they are issues that are not specific to modelling. What it means for a claim to be true about a concrete model like the model ship is just an instance of the problem of what it means for claim to be true about an object in the world. How we learn about things like the model ship is an instance of the question of how we learn about concrete objects in general. Finally, under what conditions two concrete models are identical is the question under what conditions two concrete objects are identical. These questions have been discussed extensively in philosophy, and we refer the reader to the relevant bodies of literature. We enter less clearly charted territory when we focus on models like the bridge jump model.

4.7 Non-concrete Models

The concrete models we have discussed so far are material objects: blocks of wood, hydraulic systems, and so on. But some models are non-concrete in the sense that they are not material objects. The bridge jump model we encountered in Section 1.1 is a case in point. To plan the stunt, you don't produce a scale model of the bridge and have small remote-controlled model cars jump across it (although you could!). Rather, you imagine a scenario with two inclined planes and a perfect sphere moving on them, and describe this scenario using Equations (1) and (2). The mathematized imagined scenario is your model. The Model Question now pushes us to say what that thing is that we imagine and call the model. Or in other words, what is it that we put into the box (in Figure 6) labelled 'model object' once it's not a concrete object? Saying that it is a non-concrete object is as informative as saying that a certain animal is a non-elephant. To answer the Model Question we need to move beyond this merely negative characterisation and (a) say what sort of things non-concrete model objects are; (b) formulate DEKI for such objects; and (c) show how the conditions pertaining to the Model Question are answered. This has given rise to a large discussion about the 'ontology of models'. It is impossible to review this discussion here and to formulate the different positions with precision because this would lead us deep into complex metaphysical issues that we just don't have the space to cover. Instead, we will give an intuitive sketch of two main positions in the debate, the fiction view and the structuralist view of models, and we refer the reader to the relevant literature for details.[61]

The core of the fiction view is, as Godfrey-Smith puts it, that 'modelers often take themselves to be describing imaginary biological populations, imaginary neural networks, or imaginary economies', and that these 'might be treated as similar to something that we are all familiar with, the imagined objects of literary fiction' such as 'Sherlock Holmes' London, and Tolkien's Middle Earth' (2006, 735). Drawing a parallel between models and imaginary objects has a long pedigree, also among scientists. Maxwell studied in detail the motion of an 'imaginary fluid', 'merely a collection of imaginary properties', to derive his equations of the electromagnetic field (Niven 1965, 159–60), and Einstein (1920/1999, 60) asked his readers to first 'imagine a large portion of empty space' and then 'imagine a spacious chest resembling a room with an observer inside' located in this portion of empty space, to support the equivalence principle (postulating the equivalence of gravitational and inertial mass). Imaginary fluids and observers in empty space, and indeed perfect spheres on inclined planes, are, from an ontological point of view, the same sort of things as

---

[61] For discussion of these two views, see Frigg and Nguyen (2020, Chs. 5 and 6) and references therein. For broader overviews of the ontology of models, see Frigg (2022, Ch. 14) and Gelfert (2017).

Sherlock Holmes and Middle Earth. This is the point of departure of the fiction view of models.[62]

The fiction view then says that what goes into the box labelled 'model object' is a fictional object. Views on the nature of fictional objects, or on whether fictions are objects at all, vary widely.[63] What matters in the current context is that any view on the subject matter will have to provide answers to our conditions. There is right and wrong in a fictional scenario. It's true that Sherlock Holmes is detective; it's wrong that he teaches yoga. Any account of fiction should explain what grounds these judgements, and in doing so it will give an answer to the Model-Truth Condition. Likewise, any account of fiction will have to say how we come to know things that are true in a fiction and under what conditions two fictional characters are identical. In doing so, the account will answer the Model-Epistemology Condition and the Model-Identity Condition. One can then swap concrete objects for fictional objects in the DEKI account, and once we've done that we've met the three conditions that pertain to the Model Question. One can then also attribute features to fictional objects, such objects can be said to have those features, and once they can have features, we can interpret features, and interpreted features can be exemplified. This is all we need, and the rest of the DEKI apparatus, in particular keys and imputation remain unchanged. In sum, the fiction view accounts for the use of non-concrete models by swapping concrete objects for fictional objects, which allows it to keep using the rest of the DEKI account unaltered.

The structuralist view takes a different route and takes as its leading idea the fact that models are typically given a mathematical formulation (recall the use of Newtonian mechanics in Section 1.1), rather than the fact that their construction seems to involve imagination. As we have seen in Section 2, a common way of analysing the use of mathematics in science is structuralist: mathematics describes structures and structures serve as models. So the structuralist move is to replace concrete objects with mathematical structures, in the case of the bridge jump model the structure consisting of the real numbers with the trajectory specified by Equation (2) defined upon them, and have such structures take the place of the model object in the DEKI scheme. You may now wonder what happened to the 'physical content' of a model. The trajectory in the bridge jump model is a *trajectory*, not a bare mathematical function. But this is hard to make sense of when considering a mathematical structure. The idea here is that the non-mathematical content of the model enters through the interpretation.

---

[62] Of course, this way of speaking doesn't imply that imagination in science works exactly the same way as it does in other contexts. For further discussions, see Stuart (2018), French (2020), Murphy (2020), and Salis and Frigg (2020).

[63] For surveys see, for instance, Friend (2007), Kroon and Voltolini (2018), and Salis (2013).

Endowing a structure with an interpretation turns it into a Z-representation, and hence a model. As in the fiction case, once we have a Z-representation in place, it doesn't matter whether that representation is a concrete or an abstract object and the DEKI conditions apply unaltered. In sum, according to the structuralist view a model is an interpreted structure.

At this point we're not taking sides with one or the other approach. Each has its pros and cons, but these largely have to do with issues other than those that arise in connection with DEKI. As far as DEKI is concerned, both are viable options.

4.8 Limit Keys for Mechanical Models

In the case of the ship we used a key based on Equation (7) to convert features of the model into features to be imputed to the target. What key do we use when converting features of the bridge jump model into features of the target? An important way of thinking about mechanical models like the bridge jump model is to understand their relation to their target in terms of limits: the model is seen has having features that are 'extremal' versions of the target features. Model surfaces are frictionless, while target surfaces are slippery; model planets are spherical, while real planets are roundish; and so on. A key for such features can be formulated by exploiting how mathematical limits work.

Let us briefly introduce the main idea of a limit. Consider a function $f(x)$, where x is a real number, and ask how this function behaves if we consider values for x that get ever closer to a particular value a. This amounts to taking a *limit*, and we write $\lim_{x \to a} f(x)$ to express this. The limit expresses how a function behaves if the value of x tends to a particular value. If the limit exists, this means that the closer x gets to a, the closer the values of $f(x)$ get to the limit value. Intuitively, this means that one can keep $f(x)$ as close to the limit value as one wants by keeping x sufficiently close to a. As an example consider the function $f(x) = 1/(1 + x)$ and ask how the function behaves as $x \to 0$. It's obvious that $\lim_{x \to 0} f(x) = 1$, which means that the closer x gets to 0, the closer the values of $1/(1 + x)$ stay to 1.

It's crucial to bear mind that the limit reflects how the function behaves as x gets closer and closer (and indeed arbitrarily close) to a, but without ever reaching it. So $\lim_{x \to a} f(x)$ is not the same as $f(a)$, the *value at the limit*. In our simple example it so happens that $\lim_{x \to 0} f(x) = f(0)$. If it is the case that the limit value and the value at the limit are identical, then the limit is *regular*. If the two values are different (or if one doesn't exist), then the limit is singular.[64] As an example, consider the function $g(x)$ that assumes the value 1 for all $x \neq 0$

---

[64] For a discussion of the philosophical significance of singular limits, see Butterfield (2011).

and the value 2 for $x = 0$. Here, $\lim_{x \to 0} g(x) = 1$ and $g(0) = 2$; hence, $\lim_{x \to 0} g(x) \neq g(0)$. This limit is singular.

We can now generalise the idea of limits and consider the limit of models rather than functions. This is motivated by the fact that models often have parameters that we want to vary. In the bridge jump stunt, for instance, the real car experiences air resistance, which isn't present in the model. We now introduce the parameter r for air resistance and consider a model (or collection of models) $M(r)$ that depends on the value of r like a function depends on x. The model we considered in Section 1.1 was one with zero air resistance, which corresponds to $M(0)$. A model that is an accurate representation of the target (with respect to air resistance) has a value for r that is greater than 0. This allows us to reformulate the question of what the model in Section 1.1 tells us about movement of the real car (at least with respect to the influence of air resistance) as the question of how the relevant features of $M(0)$ relate to the relevant features of $M(r)$ for some $r > 0$.

The relevant feature is the car's trajectory ϕ. Now let $\phi(0)$ be the trajectory in $M(0)$ (specified by Equation (2)) and $\phi(r)$ the trajectory in $M(r)$. What we have learned about limits gives us an answer to our question: if $\lim_{r \to 0} \phi(r)$ exists and is regular, then we know that the closer r gets to zero, the closer the behaviour of $M(r)$ gets to the behaviour of $M(0)$, meaning that the closer r gets to zero, the closer the trajectory $\phi(r)$ approaches the trajectory $\phi(0)$. Hence, if we know (a) that the limit exists and is regular, (b) how $M(0)$ behaves, and (c) that the real value of r is (in some sense that's relevant to the problem at hand) close to zero, then we know that the behaviour of $M(r)$ is close to the behaviour of $M(0)$, which means that $M(0)$ provides information about $M(r)$. If someone reasons in this way, they use what we call a *limit key*.[65]

As with many conditionals, the crux lies in the antecedent: even if we are able to tell how $M(0)$ behaves via the model analysis described in Section 1.1 and to estimate the value of r in the target, how do we know that the limit exists and is regular? This is a question that has to be settled on a case-by-case basis. In our example of the car jump, it is possible to explicitly construct both a model $M(0)$ for the motion in the vacuum and a model $M(r)$ for the motion with air

---

[65] We develop this line of reasoning in more detail in Nguyen and Frigg (2020). It's worth noting that our formulation of limit-based reasoning in terms of model behaviour and features being 'close to' target behaviour and features relies on a more general principle: limit-based reasoning, when successful, tells us that, for every value of ϵ, there is some value δ such that if the model feature is no more than δ away from the target feature, then the model behaviour is no more than ϵ away from the target behaviour. This means that a limit-based key allows us to reason as follows: if we know that the target feature is no more than δ away from the model feature, then we can infer that the target behaviour will be no more than ϵ away from the model behaviour.

resistance, and one can then prove mathematically that the limit of $\phi(r)$ for $r\to0$ exists and is equal to $\phi(0)$: $\lim_{r\to0}\phi(r) = \phi(0)$.[66] This shows that the limit is regular and that a limit key can be used. In our case this means that as long as the air resistance is relatively small, which is the case for a car that moves through the air at something like 50 km/h, the trajectory of the car with air resistance is close to the trajectory of a car that moves without air resistance. We know the trajectory of the car without air resistance (that's what our model gives us!), and so we can infer that the real car's trajectory is close to our model trajectory. This, we submit, is the reasoning you employ when you plan the stunt with your model.

Unfortunately, this is not the end of the story. The argument in the previous paragraph only concerns air resistance, but we have made a whole lot of further idealisations, for instance, that the car is spherical and that the leaves of the bridge are perfect planes. These are not negligible, and they would have to be dealt with in the same way as air resistance: one would have to introduce parameters for the shape of the car and the surface of the bridge, and an expression for air resistance for non-spherical objects would have to be found. One would then have to run through the above procedure for *all* these parameters and show that the relevant limits are regular. This is unrealistic. Knowing all relevant factors would require a God's eye perspective that mortal scientists don't have.

Does this pull the rug from underneath limiting reasoning? For those who require mathematical proofs it does. But in the empirical sciences one mostly has to make do with less rigor. In situations like these, scientists will appeal to their background knowledge about forces and the behaviour of objects, and they will form a qualitative judgement concerning whether the model results are close the behaviour of the target. Has this made limits and limit reasoning obsolete? No. In making these judgements scientists in effect engage in a thought experiment in which an omniscient creature first writes down the perfect model of the situation and then takes all the relevant limits. That this exercise would yield the desired result – that the relevant limits exist and are regular – is then a transcendental assumption for the use of such models. If one of the limits were singular, the model behaviour could be radically different from the behaviour of the target, which would undercut the use of limit

---

[66] This requires some work. The broad outlines of the task are as follows. First, formulate the equation of motion for the sphere in the vacuum, Equation (1), which is the mathematical expression of $M(0)$. Solve this equation under the relevant boundary conditions. This yields $\phi(0)$, Equation (2), which reflects the relevant behaviour of $M(0)$. Then formulate the equation of motion for the sphere moving with air resistance, assuming air resistance is given by Stokes' law. This is the mathematical expression of $M(r)$ for $r > 0$. This yields an ordinary inhomogeneous second-order differential equation. Solve this equation and use the boundary conditions to determine its integration constants. This yields $\phi(r)$, which reflects the relevant behaviour of $M(r)$. Using Taylor expansions one can then prove that $\lim_{r\to0}\phi(r)$ exists and is equal to $\phi(0)$.

reasoning. But mechanical models of the sort we have been discussing are widely used with great success, which would seem to show that the transcendental assumption is sound in many contexts.

4.9 Coda

In this section, we've seen two different keys in action: the key accompanying the concrete ship model, which relied on a background theory and empirical observations concerning how resistance forces scale with size, and the key accompanying the bridge jump model, which relied on taking certain features of the target system to extremal values and assuming that this limiting process coincided with the limit behaviour of the model.

We can also use the DEKI framework to further our understanding of how we use other kinds of epistemic representations to learn about the world. Recall again the keys that accompany the tube map and the factual images. These keys associate colour features with non-colour features of the targets of these visual representations.[67] We submit that many visual representations work this way, and the fact that DEKI captures how they work suggests, as we argue in detail elsewhere (Frigg and Nguyen 2019), that the account can also be utilised to help us understand how representations work outside of the scientific context. Consider again Philip Jackson's statue of Mahatma Gandhi in London's Parliament Square. The statue exemplifies certain features: its garb is humble; the statue's face is cast in a serene, quietly confident, expression; and the statue's plinth is lower than that of the other statues in the square. The key acts as the identity on the first two of these features, and they are imputed to Ghandi unchanged. This explains how the statue represents Ghandi's garb and facial expression. But the representational role of the third feature involves some symbolism; the key acts to take the plinth's height to something different: Ghandi's self-identification as a man of the people. In this way DEKI helps us understand how this symbolism works. The key connects a seemingly insignificant feature of the statue considered as a concrete object with a feature of central importance to the way it represents its subject.[68]

---

[67] In the case of the tube map, the key also relies on a proposed homeomorphism, a mapping that preserves topological structure, between the map and the underground system. This helps us understand what accounts like PROPOSED RESEMBLANCE get right: sometimes the keys associated with epistemic representations do rely on proposed resemblance relations, and this can be captured by DEKI.

[68] Different contextual considerations may also yield different readings of the work. The fact that the statue is positioned in Parliament square, outside an institution with a significant colonial history, may, in a certain context, be another exemplified feature of the work, highlighting a tension in the way contemporary politicians celebrate the legacy of the 'mother of all parliaments'. We discuss the 'flexibility' of artistic interpretation in Frigg and Nguyen (2019).

DEKI's application is not restricted to these more obvious cases of 'representational art'. It also allows us to understand how abstract works of art like Fischli and Weiss' *Der Lauf der Dinge* (in English: 'The Way Things Go') film of a Rube Goldberg machine can provide existential insight into the human condition, consisting as it does of a sequence of carefully calibrated but ultimately aimless events. The film provides access to the features it exemplifies, and the key associated with the work acts to transform these features into the ones we ultimately attribute to the film's subject matter (Frigg and Nguyen 2019).

Thus, we submit, DEKI is not restricted to the scientific context; it provides a framework in which to understand how epistemic representations (models, visual image, or works of art) more generally work. And in each of these cases, such an understanding requires knowing how to reason with its key, which in turn requires understanding the practices, conventions, and background knowledge of those who construct and reason with the representation. Without this, we fail to understand how the representations work: a misinformed shipbuilder simply scales the resistance faced by the model ship; a naïve user of the tube map mistakenly assumes it represents distance between stops; and an unsuspecting bystander in Parliament Square fails to notice the height of the statue's plinth.

What holds for keys holds for the other conditions too: how scientists and artists create or turn objects (concrete or non-concrete) into Z-representations, how they establish denotation relations between Z-representations and their targets (if any), and how 'context' operates to turn a feature (I-)instantiated by such an object into an exemplified feature all depend on the contexts in which those representations are embedded. In general, the DEKI conditions are stated at an abstract level, and they need to be concretized in every particular case, where the details of the concretization will depend on the details of the practices associated with the representation.

So DEKI can be used to help understand epistemic representation in general. But even if we restrict ourselves to the scientific context, the abstract aspect of DEKI opens up novel theoretical avenues to philosophers of science investigating model-based science (and to scientists hoping to understand how their models relate to the world): if we're right, then one won't understand how a model works by investigating it in isolation from its use. Understanding the latter is an essential aspect of understanding how the model functions as a representation. Thus, DEKI can act to structure investigations associated with what is known as the 'philosophy of science in practice' movement.[69] By combining the analytical framework offered by DEKI, which is skeletal by

---

[69] For more on this movement, see https://philosophy-science-practice.org/, and the mission statement offered there.

design, with rich historical, philosophical, and sociological inquiry into the details of contexts in which scientists construct and reason with their models, we do justice to the ingenious ways in which they generate knowledge and understanding about the natural and social worlds. Another book in this series does this by investigating the use of model organisms through the DEKI lens (Ankeny and Leonelli 2021). We hope there is more to come.

References

Ankeny, R. A., & Leonelli, S. (2021). *Model organisms* (Elements in the Philosophy of Biology). Cambridge: Cambridge University Press.

Argyris, J. H., Faust, G., & Haase, M. (1994). *An exploration of chaos. An introduction for natural scientists and engineers*. Amsterdam: North-Holland.

Baxter, R. J. (1982). *Exactly solved models in statistical mechanics*. London: Academic Press.

Black, M. (1973). How do pictures represent? In E. Gombrich, J. Hochberg, & M. Black (Eds.), *Art, perception, and reality* (pp. 95–130). Baltimore and London: Johns Hopkins University Press.

Boesch, B. (2017). There is a special problem of scientific representation. *Philosophy of Science, 84*(5), 970–81.

Bogen, J., & Woodward, J. (1988). Saving the phenomena. *Philosophical Review, 97*(3), 303–52.

Bolinska, A. (2013). Epistemic representation, informativeness and the aim of faithful representation. *Synthese, 190*(2), 219–34.

Brading, K., & Landry, E. (2006). Scientific structuralism: Presentation and representation. *Philosophy of Science, 73*(5), 571–81.

Brandom, R. B. (1994). *Making it explicit: Reasoning, representing and discursive commitment*. Cambridge, MA: Harvard University Press.

Brandom, R. B. (2000). *Articulating reasons: An introduction to inferentialism*. Cambridge, MA: Harvard University Press.

Bueno, O. (2010). Models and scientific representations. In P. D. Magnus, & J. Busch (Eds.), *New waves in philosophy of science* (pp. 94–111). Hampshire: Palgrave MacMillan.

Bueno, O., & French, S. (2011). How theories represent. *The British Journal for the Philosophy of Science, 62*(4), 857–894.

Butterfield, J. (2011). Less is different: emergence and reduction reconciled. *Foundations of Physics, 41*, 1065–135.

Callender, C., & Cohen, J. (2006). There is no special problem about scientific representation. *Theoria, 21*(55), 7–25.

Carlton, J. S. (2007). *Marine propellers and propulsion*. Oxford: Butterworth-Heineman.

Chakravartty, A. (2010). Informational versus functional theories of scientific representation. *Synthese, 172*(2), 197–213.

Contessa, G. (2007). Scientific representation, interpretation, and surrogative reasoning. *Philosophy of Science, 74*(1), 48–68.

de Donato Rodriguez, X., & Zamora Bonilla, J. (2009). Credibility, idealisation, and model building: An inferential approach. *Erkenntnis, 70*(1), 101–18.

Decock, L., & Douven, I. (2011). Similarity after Goodman. *Review of Philosophy and Psychology, 2*(1), 61–75.

Díez, J. (2020). An ensemble-plus-standing-for account of scientific representation: no need for (unnecessary) abstract objects. In C. Martínez-Vidal, & J. L. Falguera (Eds.), *Abstract objects. For and against* (pp. 133–49). Cham: Springer.

Ducheyne, S. (2012). Scientific representations as limiting cases. *Erkenntnis, 76* (1), 73–89.

Einstein, A. (1920/1999). *Relativity: The special and general theory.* London: Methuen.

Elgin, C. Z. (1983). *With reference to reference.* Indianapolis and Cambridge: Hackett.

Elgin, C. Z. (2010). Telling instances. In R. Frigg, & M. C. Hunter (Eds.), *Beyond mimesis and convention: Representation in art and science* (pp. 1–18). Berlin and New York: Springer.

Elgin, C. Z. (2017). *True enough.* Cambridge, MA: MIT Press.

French, S. (2020). Imagination in scientific practice. *European Journal for Philosophy of Science, 10*(3), 1–19.

French, S. (2021). Identity conditions, idealisations and isomorphisms: A defence of the Semantic Approach. *Synthese, 198*, 5897–917.

French, S., & Ladyman, J. (1999). Reinflating the semantic approach. *International Studies in the Philosophy of Science, 13*, 103–21.

Friend, S. (2007). Fictional characters. *Philosophy Compass, 2*(2), 141–56.

Frigg, R. (2006). Scientific representation and the semantic view of theories. *Theoria, 55*(1), 49–65.

Frigg, R. (2022). *Models and theories.* London: Routledge (forthcoming).

Frigg, R., & Nguyen, J. (2017). Scientific representation is representation as. In H.-K. Chao, & R. Julian (Eds.), *Philosophy of science in practice: Nancy Cartwright and the nature of scientific reasoning* (pp. 149–79). Cham: Springer.

Frigg, R., & Nguyen, J. (2019). Of barrels and pipes: representation-as in art and science. In S. Wuppuluri (Ed.), *On art and science. Tango of an eternally inseparable duo* (pp. 181–202). Cham: Springer.

Frigg, R., & Nguyen, J. (2020). *Modelling nature. An opinionated introduction to scientific representation.* Berlin and New York: Springer.

Frigg, R., & Votsis, I. (2011). Everything you always wanted to know about structural realism but were afraid to ask. *European Journal for Philosophy of Science, 1*(2), 227–76.

Gelfert, A. (2017). The ontology of models. In L. Magnani, & T. Bertolotti (Eds.), *Springer handbook of model-based science* (pp. 5–23). Dordrecht Heidelberg: Springer.

Giere, R. N. (1988). *Explaining science: A cognitive approach.* Chicago and London: University of Chicago Press.

Giere, R. N. (2004). How models are used to represent reality. *Philosophy of Science, 71*(4), 742–52.

Giere, R. N. (2010). An agent-based conception of models and scientific representation. *Synthese, 172*(1), 269–81.

Godfrey-Smith, P. (2006). The strategy of model-based science. *Biology and Philosophy, 21*(5), 725–40.

Goodman, N. (1972). Seven strictures on similarity. In N. Goodman (Ed.), *Problems and projects* (pp. 437–46). Indianapolis and New York: Bobbs-Merrill.

Goodman, N. (1976). *Languages of art* (2nd ed.). Indianapolis and Cambridge: Hackett.

Hacking, I. (1983). *Representing and intervening: Introductory topics in the philosophy of natural science.* Cambridge: Cambridge University Press.

Hartmann, S. (1995). Models as a tool for theory construction: Some strategies of preliminary physics. In W. E. Herfel, W. Krajewski, I. Niiniluoto, & R. Wojcicki (Eds.), *Theories and models in scientific processes (Poznan Studies in the Philosophy of Science and the Humanities 44)* (pp. 49–67). Amsterdam and Atlanta: Rodopi.

Hesse, M. (1963). *Models and analogies in science.* London: Sheed and Ward.

Hodges, W. (1997). *A shorter model theory.* Cambridge: Cambridge University Press.

Hughes, R. I. G. (1997). Models and representation. *Philosophy of Science, 64*, S325–S336.

Khalifa, K., Millson, J., & Risjord, M. (2022). Scientific representation: An inferentialist-expressivist Manifesto. In K. Khalifa, I. Lawler, & E. Shech (Eds.), *Scientific understanding and representation: Modeling in the physical sciences* (pp. TBC). TBC: Routledge.

Khosrowi, D. (2020). Getting serious about shared features. *The British Journal for the Philosophy of Science, 71*(2), 523–46.

Kirkham, R. L. (1992). *Theories of truth: a critical introduction.* Cambridge, MA: MIT Press.

Knuuttila, T. (2021. Imagination extended and embedded: artifactual versus fictional accounts of models. *Synthese, 198*, 5077–97.

Kroon, F., & Voltolini, A. (2018). Fictional entities. In E. N. Zalta (Ed.), *The Stanford encyclopedia of philosophy,* https://plato.stanford.edu/archives/win2018/entries/fictional-entities/.

Kuhn, T. S. (1957). *The Copernican Revolution. Planetary astronomy in the development of Western thought* (2nd ed.). Massachusetts: Harvard University Press.

Kulvicki, J. (2006). Pictorial representation. *Philosophy Compass, 1*(6), 535–46.

Kuorikoski, J., & Lehtinen, A. (2009). Incredible worlds, credible results. *Erkenntnis, 70*(1), 119–31.

Levy, A. (2015). Modeling without models. *Philosophical Studies, 152*(3), 781–98.

McCullough-Benner, C. (2020). Representing the world with inconsistent · mathematics. *The British Journal for the Philosophy of Science, 71*(4), 1331–58.

Molland, A. F. (Ed.). (2008). *The maritime engineering reference book. A guide to ship design, construction and operation.* Oxford: Butterworth-Heineman.

Morgan, M. (2012). *The world in the model. How economists work and think.* Cambridge: Cambridge University Press.

Morgan, M., & Morrison, M. (Eds.). (1999). *Models as mediators: Perspectives on natural and social science.* Cambridge: Cambridge University Press.

Murphy, A. (2020). Towards a pluralist account of the imagination in science. *Philosophy of Science, 87*(5), 957–67.

Murzi, J., & Steinberger, F. (2017). Inferentialism. In B. Hale, C. Wright, & A. Miller (Eds.), *A companion to the philosophy of language* (2nd ed., Vol. 1). Chichester: Wiley Blackwell.

Nguyen, J. (2016). On the pragmatic equivalence between representing data and phenomena. *Philosophy of Science, 83*(2), 171–91.

Nguyen, J. (2020). It's not a game: accurate representation with toy models. *The British Journal for the Philosophy of Science, 71*(3), 1013–41.

Nguyen, J., & Frigg, R. (2021). Mathematics is not the only language in the book of nature. *Synthese, 198*, 5941–5962.

Nguyen, J., & Frigg, R. (2020). Unlocking limits. *Argumenta, 6*(1), 31–45.

Nguyen, J., & Frigg, R. (2022). Maps, models, and representation. In K. Khalifa, I. Lawler, & E. Shech (Eds.), *Scientific understanding and representation: Modeling in the physical sciences* (pp. TBC). TBC: Routledge.

Niiniluoto, I. (1988). Analogy and similarity in scientific reasoning. In D. H. Helman (Ed.), *Analogical reasoning: Perspectives of artificial intelligence, cognitive science, and philosophy* (pp. 271–98). Dordrecht: Kluwer.

Niven, W. D. (1965). *The scientific papers of James Clerk Maxwell.* New York: Dover.

Norton, J. (2008). The dome: An unexpectedly simple failure of determinism. *Philosophy of Science, 75*(5), 786–98.

Parker, W. (2015). Getting (even more) serious about similarity. *Biology & Philosophy, 30,* 267–76

Pero, F., & Suárez, M. (2016). Varieties of misrepresentation and homomorphism. *European Journal for Philosophy of Science, 6*(1), 71–90.

Pincock, C. (2005). Overextending partial structures: Idealization and abstraction. *Philosophy of Science, 72*(5), 1248–59.

Pincock, C. (2012). *Mathematics and scientific representation.* Oxford: Oxford University Press.

Pincock, C. (2022). Concrete Scale Models, Essential Idealization, and Causal Explanation. *The British Journal for the Philosophy of Science, 73*(2), 299–323

Potochnik, A. (2017). *Idealization and the aims of science.* Chicago and London: University of Chicago Press.

Putnam, H. (1981). *Reason, truth, and history.* Cambridge: Cambridge University Press.

Quine, W. V. O. (1969). *Ontological relativity and other essays.* New York: Columbia University Press.

Russell, B. (1919/1993). *Introduction to mathematical philosophy.* London and New York: Routledge.

Ruyant, Q. (2021). True Griceanism: Filling the gaps in Callender and Cohen's account of scientific representation. *Philosophy of Science, 88*(3), 533–53.

Salis, F. (2013). Fictional entities. In J. Branquinho, & R. Santos (Eds.), *Online companion to problems in analytical philosophy.* http://compendioemlinha.letras.ulisboa.pt.

Salis, F., & Frigg, R. (2020). Capturing the scientific imagination. In P. Godfrey-Smith, & A. Levy (Eds.), *The scientific imagination. Philosophical and psychological perspectives* (pp. 17–50). New York: Oxford University Press.

Salis, F., Frigg, R., & Nguyen, J. (2020). Models and denotation. In C. Martínez-Vidal, & J. L. Falguera (Eds.), *Abstract objects: For and against* (pp. 197–219). Cham: Springer.

Shapiro, S. (1983). Mathematics and reality. *Philosophy of Science, 50*(4), 523–48.

Shech, E. (2015). Scientific misrepresentation and guides to ontology: The need for representational code and contents. *Synthese, 192,* 3463–85.

Slutkin, G. (2013). Violence is a contagious disease. In S. M. Patel DM, R. A. Taylor (Eds.), *Contagion of violence: Workshop summary* (pp. 94–111). Washington, DC: National Academies Press.

Sterrett, S. G. (2009). Similarity and dimensional analysis. In A. Mejers (Ed.), *Philosophy of technology and engineering sciences* (pp. 799–823). Amsterdam: Elsevier/North Holland.

Sterrett, S. G. (2017a). Experimentation on analogue models. In L. Magnani, & T. Bertolotti (Eds.), *Springer handbook of model-based science* (pp. 857–78). Cham: Springer.

Sterrett, S. G. (2017b). Physically similar systems – A history of the concept. In L. Magnani, & T. Bertolotti (Eds.), *Springer handbook of model-based science* (pp. 377–411). Cham: Springer.

Sterrett, S. G. (2020). Scale modeling. In D. Michelfelder, & N. Doorn (Eds.), *Routledge handbook of philosophy of engineering* (pp. Ch. 29). London: Routledge. https://doi.org/10.4324/9781315276502

Stuart, M. T. (2018). Thought experiments: The state of the art. In M. Stuart, Y. Fehige, & J. Brown (Eds.), *The Routledge companion to thought experiments* (pp. 1–28). London: Routledge.

Suárez, M. (2003). Scientific representation: Against similarity and isomorphism. *International Studies in the Philosophy of Science, 17*(3), 225–44.

Suárez, M. (2004). An inferential conception of scientific representation. *Philosophy of Science, 71*(5), 767–79.

Suárez, M. (2015). Deflationary representation, inference, and practice. *Studies in History and Philosophy of Science, 49*, 36–47.

Suárez, M., & Solé, A. (2006). On the analogy between cognitive representation and truth. *Theoria, 55*(1), 39–48.

Swoyer, C. (1991). Structural representation and surrogative reasoning. *Synthese, 87*(3), 449–508.

Tegmark, M. (2008). The mathematical universe. *Foundations of Physics, 38* (2), 101–50.

Teller, P. (2001). Twilight of the perfect model model. *Erkenntnis, 55*(3), 393–415.

Thomasson, A. L. (2020). If models were fictions, then what would they be? In A. Levy, & P. Godfrey-Smith (Eds.), *The scientific imagination. Philosophical and psychological perspectives* (pp. 51–74). New York: Oxford University Press.

Thomson-Jones, M. (2010). Missing systems and face value practice. *Synthese, 172*(2), 283–99.

Toon, A. (2012). *Models as make-believe. Imagination, fiction and scientific representation*. Basingstoke: Palgrave Macmillan.

Tversky, A. (1977). Features of similarity. *Psychological Review, 84*(4), 327–52.

van Fraassen, B. C. (1980). *The scientific image*. Oxford: Oxford University Press.

van Fraassen, B. C. (2008). *Scientific representation: Paradoxes of perspective.* Oxford: Clarendon Press.

Weisberg, M. (2007). Who is a modeler? *The British Journal for the Philosophy of Science, 58*(2), 207–33.

Weisberg, M. (2012). Getting serious about similarity. *Philosophy of Science, 79*(5), 785–94.

Weisberg, M. (2013). *Simulation and similarity: Using models to understand the world.* Oxford: Oxford University Press.

Weisberg, M. (2015). Biology and philosophy symposium on simulation and similarity: Using models to understand the world: Response to critics. *Biology and Philosophy, 30*(2), 299–310.

Wiley S. A., Levy, M. Z., & Branas C. C. (2016). The impact of violence interruption on the diffusion of violence: A mathematical modeling approach. In G. Letzer. et. al. (Eds.), *Advances in the mathematical sciences* (Vol. 6, pp. 225–49, Association for Women in Mathematics Series). Cham: Springer.

Xia, Z. (1992). The existence of noncollision singularities in Newtonian systems. *Annals of Mathematics, 135*(3), 411–68.

Acknowledgements

The content of this Element has been presented at numerous events, and discussed with many friends and colleagues (too many to be named here) during the better part of the past decade. We are grateful for all the wonderful feedback and support we have received. Thanks to Jacob Stegenga for inviting us to participate in this series, and to two anonymous referees for detailed and helpful comments. JN thanks the Jacobsen Fund and the Jeffrey Rubinoff Sculpture Park for postdoctoral support whilst working on this project. He also thanks Ashton and Yakobi for providing some much-needed perspective during the global pandemic through which this element was written. RF thanks Benedetta for her unwavering support.

Cambridge Elements ☰

Philosophy of Science

Jacob Stegenga

University of Cambridge

Jacob Stegenga is a Reader in the Department of History and Philosophy of Science at the University of Cambridge. He has published widely on fundamental topics in reasoning and rationality and philosophical problems in medicine and biology. Prior to joining Cambridge he taught in the United States and Canada, and he received his PhD from the University of California San Diego.

About the Series

This series of Elements in Philosophy of Science provides an extensive overview of the themes, topics, and debates which constitute the philosophy of science. Distinguished specialists provide an up-to-date summary of the results of current research on their topics, as well as offering their own take on those topics and drawing original conclusions.

Cambridge Elements ≡

Philosophy of Science

Elements in the Series

A full series listing is available at: www.cambridge.org/EPSC

Printed in the United States
by Baker & Taylor Publisher Services